AMERICAN MATHEMATICAL SOCIETY
TRANSLATIONS
Series 2
Volume 116

Three Papers on Dynamical Systems

by

A. G. KUŠNIRENKO
A. B. KATOK
V. M. ALEKSEEV

AMERICAN MATHEMATICAL SOCIETY
PROVIDENCE, RHODE ISLAND
1981

Edited by J. SZŰCS

1980 *Mathematics Subject Classifications.* Primary 34C15, 34F05, 58F15, 70F15; Secondary 58F18, 58F20, 58F22, 58F25, 70F07.

Library of Congress Cataloging in Publication Data

Main entry under title:

Three papers on dynamical systems.

(American Mathematical Society translations; ser. 2, v. 116)

Bibliography: p.

Contents: Problems in the general theory of dynamical systems on a manifold / A. G. Kusnirenko–Dynamical systems with hyperbolic structure / A. B. Katok–Quasirandom oscillations and qualitative questions in celestial mechanics / V. M. Alekseev.

1. Differentiable dynamical systems–Addresses, essays, lectures. 2. Nonlinear oscillations–Addresses, essays, lectures. 3. Mechanics, Celestial–Addresses, essays, lectures. I. Kushnirenko, A. G. Zadachi obshcheĭ teorii dinamicheskikh sistem na mnogoobrazii. English. 1981. II. Katok, A. B. Dinamicheskie sistemy s giperbolicheskoĭ strukturoĭ. English. 1981. III. Alekseev, V. M. Kvazisluchnaĭnye kolebaniia͡ i kachestvennye voprosy nebesnoĭ mekhaniki. English. 1981. IV. Series.

QA3.A572 ser. 2, vol. 116 [QA614.8] 510s [515.3′52]

ISBN 0-8218-3066-X

81-4981

AACR2

Printed in the United States of America

TABLE OF CONTENTS

RUSSIAN TABLE OF CONTENTS*

---

* The American Mathematical Society scheme for transliteration of Cyrillic may be found at the end of index issues of *Mathematical Reviews.*

SMOOTH DYNAMICAL SYSTEMS

The literature on smooth dynamical systems is substantial. In selecting material for our lectures we have set ourselves a twofold aim. On the one hand we have tended to give a more or less connected account of a number of contemporary results associated with general problems of the classification of dynamical systems, by describing "rough" and "typical" properties, etc. On the other hand we wish to emphasize that the general constructions arising here are connected with ideas going back to the classics, and they permit us to obtain new information on qualitative properties of some long known problems.

We have not set ourselves the objective of recounting in detail the genesis and evolution of the ideas and notions under consideration, since this would require a substantial increase in size. The bibliographic citations we give do not pretend to be complete; in addition to the few remarks and references of a historical character that are presented by each author, we recommend consultation of the fundamental contribution *Geodesic flows on closed Riemannian manifolds of negative curvature* by D. V. Anosov. As far as we know, this paper gives the most detailed historical survey of everything connected with the "hyperbolic" behavior of the trajectories of dynamical systems–a property playing a basic role in the circle of questions under consideration.

V. M. Alekseev
A. B. Katok
A. G. Kušnirenko

Amer. Math. Soc. Transl.
(2) Vol. 116, 1981

PROBLEMS IN THE GENERAL THEORY OF DYNAMICAL SYSTEMS ON A MANIFOLD*

A. G. KUŠNIRENKO

CONTENTS

§1. Simplest examples of dynamical systems

1. Let M be a C^∞-manifold. By a classical dynamical system on M, we mean either a diffeomorphism $\varphi\colon M \longrightarrow M$, or a one-parameter group of diffeomorphisms φ_t of M, differentiable in t (i.e. a family of diffeomorphisms φ_t of M, differentiable with respect to t, and satisfying the condition $\varphi_{t_1+t_2} = \varphi_{t_1} \circ \varphi_{t_2}$ for any $t_1, t_2 \in \mathbf{R}$). A vector field $X = d\varphi/dt|_{t=0}$ may be associated with every such one-parameter group; and, conversely, in the case where M is compact, every smooth vector field gives rise (by a well-known theorem) to a group of diffeomorphisms of M, depending smoothly upon the parameter t. Such a group of diffeomorphisms (or sometimes the vector field which generates it) is called a *dynamical system with continuous time*, or a *flow*, while the set $\{\varphi^n\}_{n\in\mathbf{Z}}$ of all powers of some diffeomorphism φ (or just this diffeomorphism itself) is often called a *dynamical system with discrete time*. From every dynamical system $\{\varphi_t\}_{t\in\mathbf{R}}$ with continuous time, one can obtain a dynamical system $\{\varphi_{n\cdot t_0}\}_{n\in\mathbf{Z}}$ with discrete time, generated by the powers of the diffeomorphism φ_{t_0}; but not every system with discrete time can be obtained from some flow in this manner.

EXAMPLE 1. *Rotation of a circle.* $M = S^1 = \{x \pmod{2\pi}\}$, $\varphi\colon S^1 \longrightarrow S^1$, $\varphi x = x + d \pmod{2\pi}$. *Case* a) $d/2\pi$ is rational; *case* b) $d/2\pi$ is irrational.

The set $\{\varphi^n x\colon n \in \mathbf{Z}\}$ (in the case of a dynamical system with discrete time), or the set $\{\varphi_t x\colon t \in \mathbf{R}\}$ (in the case of continuous time) is called the *trajectory* or *orbit* of the point $x \in M$.

---

*Translation of Ninth Math. Summer School (Kaciveli, 1971), Izdanie Inst. Mat. Akad. Nauk Ukrain. SSR, Kiev, 1972, pp. 52–124. MR **52** #1789.

1980 *Mathematics Subject Classification.* Primary 58F15; Secondary 58F25, 58F18, 58F22.

In Example 1, case a), every orbit consists of a finite number of points (the trajectories are closed); in case b) all the orbits are infinite sets, and it is easily shown that every orbit is everywhere dense in S^1.

EXAMPLE 2. $M = T^2 = S^1 \times S^1 = \{(x_1, x_2) \pmod 1\}$. A flow on M is given by a system of ordinary differential equations: $\dot{x}_1 = 1, \dot{x}_2 = \lambda$. The corresponding one-parameter group of diffeomorphisms of M may be written down explicitly:

$$\varphi_t\begin{pmatrix} x_1 \\ x_2 \end{pmatrix} = \begin{matrix} x_1 + t \pmod 1 \\ x_2 + \lambda t \pmod 1 \end{matrix}.$$

Just as in Example 1, two cases are possible: a) λ is rational, and in this case all the trajectories are closed; b) λ is irrational, and in this case it is easily shown that every trajectory is everywhere dense in T^2.

There is a relation between Examples 1 and 2, which will be discussed below.

2. DEFINITION. The point $p \in M$ is said to be a *fixed point* of a dynamical system, provided the orbit of p coincides with p. In the case of continuous time, a fixed point is often called a *singular point*, a *stationary point*, or an *equilibrium position*.

EXAMPLE 3. $M = S^2$ = Riemann sphere = $\widetilde{\mathbf{C}}$. On $\widetilde{\mathbf{C}}\backslash\infty$, a dynamical system is given by the system of two differential equations: $\dot{z} = z$. The substitution $z = 1/V$ shows that φ_t is also smooth in the neighborhood of ∞.

It is easily seen that ∞ is a fixed point and that, for any $t \in \mathbf{R}$ and $z \in \mathbf{C}$, $\varphi_t z = e^t z$. Thus the dynamical system under consideration has two fixed points, 0 and ∞, while for any other point z_0

$$\varphi_t z_0 \longrightarrow \begin{cases} \text{“}\infty\text{”} & \text{as } t \longrightarrow +\infty, \\ 0 & \text{as } t \longrightarrow -\infty. \end{cases}$$

We note that, unlike Examples 1 and 2, φ_t does not preserve any finite regular (i.e., positive on any open set) measure on S^2. In fact, for $t > 0$ we have the strict inclusion $\varphi_t\{z: |z| > 1\} \subset \{z: |z| > 1\}$.

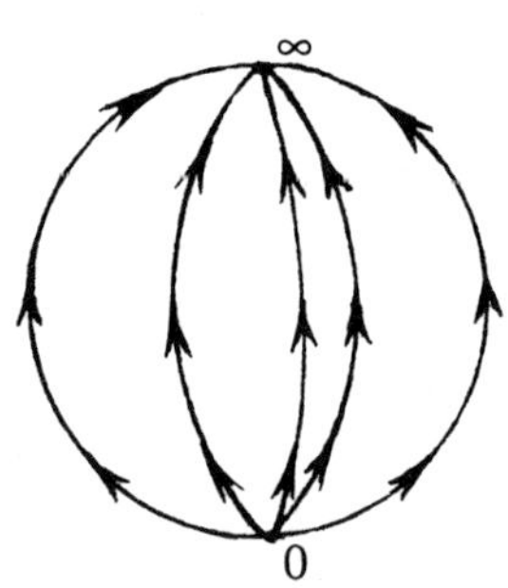

FIGURE 1

EXAMPLE 4. We take the dynamical system (with continuous time) of Example 1, and obtain from it, in the manner described above, the dynamical system (with discrete time) $\{\varphi_t: t \in \mathbf{Z}\}$. This system is generated by the diffeomorphism φ_1: $\varphi_1(\infty) = \infty$ and $\varphi_1 z = ez$.

EXAMPLE 5. $M = S^1$, $\dot{x} = \sin 3x$. In this example there are six fixed points, three attractive and three repulsive. All the remaining points form six trajectories. Every nonfixed point approaches one of the attractive fixed points as $t \longrightarrow +\infty$, and one of the repulsive fixed points as $t \longrightarrow -\infty$.

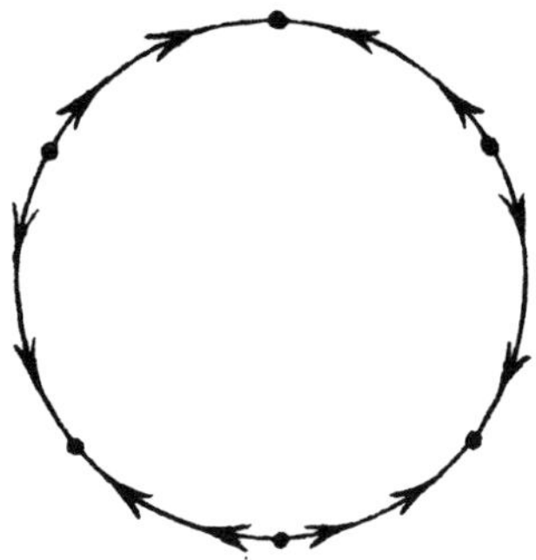

FIGURE 2

EXAMPLE 6. Let $M = S^1$ and let φ_t be the one-parameter group of diffeomorphisms of Example 5.

a) Set $\varphi = \varphi_1$.

b) Define the diffeomorphism ψ by the equality $\psi x = \varphi_1 x + 2\pi/3$.

In Example 6a), just as in Example 5, there are three attractive and three repulsive fixed points, while the closure of any other trajectory contains two fixed points, one attractive and one repulsive. In Example 6b) there are no fixed points. The orbit of the point 0 consists of three points, and likewise the orbit of π consists of three points.

DEFINITION. The point $x \in M$ is said to be a *periodic point* of the diffeomorphism φ if there exists an $n > 0$ such that $\varphi^n x = x$. The smallest such integer n is called the *period* of x.

In the case of continuous time, the trajectory of the point $x \in M$ is said to be *periodic with period* T ($T > 0$) if $\varphi_T x = x$ and $\varphi_t x \neq x$ for $0 < t < T$. The periodic trajectories of φ_t are also known as *closed* trajectories.

REMARK. Among the fixed points of the diffeomorphism φ^n are included all the periodic points of period n of the diffeomorphism φ; but, in general, not every fixed point of φ^n is periodic with period n.

Thus, in Example 6b), there are two periodic trajectories with period 3 (six periodic points). For any nonperiodic point x, $\psi^n x$ approaches one of the periodic trajectories as $n \longrightarrow +\infty$, and the other as $n \longrightarrow -\infty$.

EXAMPLE 7. As M, we take the sphere S^2, smoothly imbedded in $\mathbf{R}^3$, as shown in Figure 3 for case a), and in Figure 4 for case b). We consider the function $f = x_3|_{S^2}$ (i.e., height) on S^2, and put $-\dot{x} = \operatorname{grad} f$ ($x \in S^2$, and the metric on S^2 is induced by the imbedding in $\mathbf{R}^3$). The trajectories of these dynamical systems are known as *lines of steepest descent.*

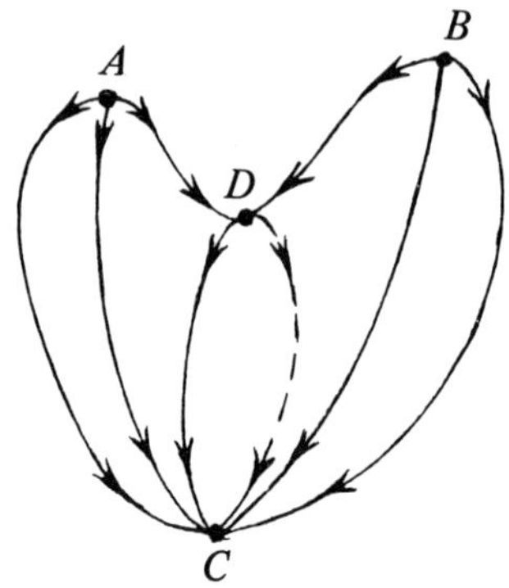

FIGURE 3

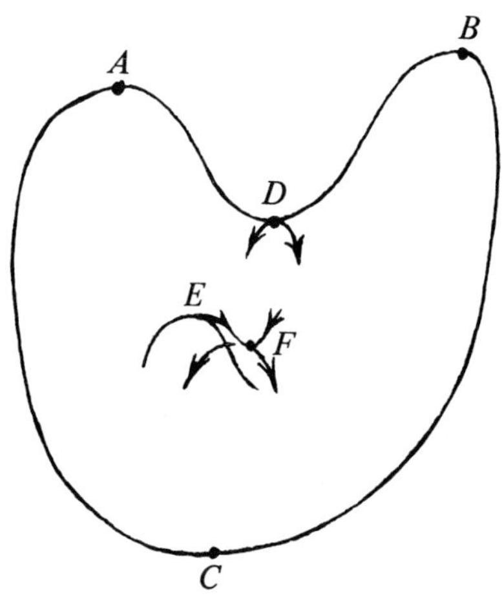

FIGURE 4

Case a). There are four fixed points, A, B, C, and D, corresponding to the critical points of the function f.

In small neighborhoods of A and B, all trajectories recede from the critical point as t increases, and in a small neighborhood of C, all trajectories approach C as t increases. D is a saddle point. There are two trajectories which approach D as $t \longrightarrow +\infty$, and two trajectories which approach D as $t \longrightarrow -\infty$. Any other trajectory bypasses the point D (i.e., leaves any small neighborhood of D as $t \longrightarrow \pm\infty$). The set consisting of those points of M which approach G as $t \longrightarrow +\infty$ is open and everywhere dense in M. This set is homeomorphic to an open disk. The complement of this set is of smaller dimension than M, and consists of the fixed points A, D, B and the two trajectories which approach D as $t \longrightarrow +\infty$. Similarly, the set consisting of those points of M which approach either A or B as $t \longrightarrow -\infty$ is open and everywhere dense in M, while the complement of this set has dimension 1.

Case b). In this example, there are six fixed points: three maximum points of f, one minimum point of f and two saddle points. The behavior of the trajectories on M

essentially depends upon how the separatrices emanating from the point D (that is, the two trajectories which approach D as $t \longrightarrow -\infty$) behave as $t \longrightarrow +\infty$.

EXERCISE. We excise from S^2 a small disk around the point C. The remaining portion of S^2 is also a disk. Sketch the trajectories of system 7b) on this disk, examining the various possible cases with respect to the behavior of the separatrices emanating from the point D.

3. In the next two examples, there naturally appear manifolds more complicated than those in the examples already considered.

EXAMPLE 8. a) We consider the sphere S^2, smoothly imbedded in $\mathbf{R}^3$, and a particle moving on the surface of S^2 without friction and with unit velocity. The phase space of this dynamical system is the manifold of unit vectors tangent to S^2; and, as is easily verified, this manifold is diffeomorphic to three-dimensional real projective space.

b) An analogous construction can be carried out for any surface $M^2 \subset \mathbf{R}^3$. The phase space of such a dynamical system will be the three-dimensional manifold of unit vectors tangent to M^2. The metric on M^2 is induced by the imbedding in $\mathbf{R}^3$, and the trajectories of the dynamical system so obtained correspond to the geodesics on the Riemannian manifold M^2.

EXAMPLE 9. Let A be a real $n \times n$ matrix. We consider the system of linear differential equations $\dot{x} = Ax$ in $\mathbf{R}^n$, and denote by φ_t the corresponding one-parameter group of linear transformations of $\mathbf{R}^n$. We fix an integer k, $1 \leqslant k \leqslant n-1$, and consider the set of all k-dimensional linear subspaces of $\mathbf{R}^n$. This set may naturally be given the structure of a smooth manifold, which is called a *real Grassmann manifold*, and is denoted by G_n^k. Since each diffeomorphism φ_t sends any k-dimensional subspace into some other k-dimensional subspace, φ_t induces a new dynamical system $\hat{\varphi}_t$ on G_n^k.

REMARK. Let the vector field X on the manifold M^n have the fixed point x_0; we wish to find (locally) the k-dimensional invariant manifolds of the flow φ_t passing through x_0. In terms of local coordinates with origin at x_0, the flow φ_t is given by the system of differential equations $\dot{x} = Ax + \cdots$. The tangent plane to any invariant k-dimensional manifold is invariant with respect to the linear system $\dot{x} = Ax$, and therefore constitutes a fixed point of the flow $\hat{\varphi}_t$ on G_n^k. Thus, the determination of the fixed points of $\hat{\varphi}_t$ in G_n^k constitutes a natural step in finding the invariant manifolds passing through x_0. In exactly the same manner, to investigate the behavior of k-dimensional manifolds which are (locally) nearly invariant, it is natural to begin by examining a small neighborhood of a fixed point of the flow $\hat{\varphi}_t$ in G_n^k.

EXERCISE. In the case where $n = 4$, $k = 2$, and A is a diagonal matrix, find the fixed points and invariant submanifolds of $\hat{\varphi}_t$. Describe the behavior of the trajectories* in the two cases of Example 7.

**Translator's note.* Presumably this refers to the trajectories of the flow $\hat{\varphi}_t$ induced by the linearized equations in the neighborhoods of the various fixed points, as explained in the preceding remark.

§2. The relation between diffeomorphisms and flows

Certain formal relations may be established between flows and diffeomorphisms.

1. Succession functions. Let the flow φ_t on an $(n + 1)$-dimensional manifold have a closed trajectory γ with period T. We take any point x_0 on γ and any n-dimensional area element P passing through x_0 and transversal to γ.

Let U be some sufficiently small neighborhood of x_0 in P. For any point $x \in U$, the trajectory $\varphi_t x$ of this point remains close to the trajectory of x_0 (that is, to γ), and consequently for some $t_0 > 0$ ($t_0 \approx T$) this trajectory will (for the first time) intersect P.

Denoting this first point of intersection by φx, we obtain a local diffeomorphism $\varphi\colon U \longrightarrow P$, where $\varphi x_0 = x_0$.

DEFINITION. By a *local diffeomorphism* of a (possibly open) manifold P, we mean a pair (U, φ) consisting of an open set $U \subset P$ and a diffeomorphism $\varphi\colon U \longrightarrow \varphi(U) \subset P$.

Let the local diffeomorphism (U, φ) of the manifold P have a fixed point x_0, and let the local diffeomorphism (U', φ') of the manifold P' have the fixed point x_0'. We say that the diffeomorphism φ at x_0 is *differentiably equivalent* to φ' at x_0' if there exists a diffeomorphism $h\colon V \longrightarrow V'$, where V and V' are open sets, $x_0 \in V \subset U$ and $x_0' \in V' \subset U'$, such that

$$h \circ \varphi|_{V \cap \varphi^{-1}(V)} = \varphi' \circ h|_{V \cap \varphi^{-1}(V)}.$$

EXERCISE. Let P' be a transversal surface element at some point $x_1 \in \gamma$ and let $\varphi'\colon U' \longrightarrow P'$ be the corresponding local diffeomorphism. Show that φ and φ' are differentiably equivalent.

It is easily seen that the study of the flow φ_t in the vicinity of γ is completely equivalent to the study of the local diffeomorphism φ. For example, the determination of the periodic trajectories of φ_t in the vicinity of γ is equivalent to the determination of the fixed and periodic points of φ, the stability of the periodic solution γ is equivalent to the stability of the fixed point x_0 of φ, and so forth.

The local diffeomorphism φ is often called a *succession function.*

2. Global succession functions. Sometimes a construction similar to the preceding may be effectively carried out in a global context. Let W^{n+1} be a compact, connected $(n + 1)$-dimensional manifold, and φ_t a flow on it. We assume that there exists a smooth compact n-dimensional submanifold $M^n \subset W^{n+1}$ having the following properties:

1) no trajectory of φ_t is tangent to M^n;

2) every trajectory has a point of intersection with M^n.

Then,

3) for any point $x \in M^n$, there exists a minimal $t_0 > 0$ such that $\varphi_{t_0} x \in M^n$. For $x \in M^n$, letting $\varphi(x)$ be the first point of intersection of $\varphi_t x$ with M^n for $t > 0$, we obtain a well-defined global "*succession diffeomorphism.*"

The manifold M^n is called a *global transversal* for the flow φ_t on W^{n+1}.

It is easily seen that the study of the flow φ_t on W^{n+1} is equivalent to the study of the diffeomorphism φ on M^n. Moreover, it can be shown that every diffeomorphism ψ which is close to φ on M^n is induced in the above-described manner by some flow ψ_t which is close to φ_t on W^{n+1}.

EXERCISE. Let M_1^n be a submanifold of W^{n+1} which is C^1-close to M^n.

a) Show that M_1^n is a global transversal for the flow φ_t.

b) Show that there exists a diffeomorphism $h: M^n \longrightarrow M_1^n$ such that the following diagram is commutative:

$$\begin{array}{ccc} M^n & \xrightarrow{\ \varphi\ } & M^n \\ h\downarrow & & \downarrow h \\ M_1^n & \xrightarrow{\ \varphi_1\ } & M_1^n \end{array}$$

(here φ_1 is the succession diffeomorphism on M_1^n).

3. Suspension. We shall now show that any diffeomorphism φ of an arbitrary manifold M^n is the succession diffeomorphism of some flow.

Consider the manifold-with-boundary $M^n \times [0, 1]$, the boundary of which consists of two copies of M^n (that is, $M^n \times 0$ and $M^n \times 1$), and consider the vector field on this manifold consisting of the unit tangent vectors in the direction of the interval $[0, 1]$. We now identify $M^n \times 0$ and $M^n \times 1$ by means of the diffeomorphism φ; that is, we regard $x \times 0$ and $\varphi x \times 1$ as one and the same point. It is easily seen that we thereby obtain a certain smooth manifold W^{n+1} and a smooth flow φ_t on it; moreover, the locus of the "pasting" operation constitutes a global transversal with succession diffeomorphism φ. The flow φ_t on W^{n+1} is called the *suspension* of the diffeomorphism φ.

EXERCISE. Show that the flow of Example 2 is the suspension of the diffeomorphism of Example 1.

REMARK. In fact, the preceding arguments contained a logical gap. In the set obtained after identifying $M^n \times 0$ and $M^n \times 1$, the structure of a smooth manifold does not just arise by itself, but must be introduced. We shall not do this, but rather mention a construction of the suspension somewhat different in form from the preceding one.

Consider the manifold $\widetilde{W}^{n+1} = M^n \times \mathbf{R}$ and the flow defined by the formula $\widetilde{\varphi}_t(x, \tau) = (x, \tau + t)$, and also the diffeomorphism $\widetilde{\varphi}$ of $\widetilde{W}^{n+1}$ defined by the formula $\widetilde{\varphi}(x, \tau) = (\varphi x, \tau - 1)$. Clearly, $\widetilde{\varphi}_t \circ \widetilde{\varphi} = \widetilde{\varphi} \circ \widetilde{\varphi}_t$ for any t. Now, we shall say that points a and b of the manifold $\widetilde{W}^{n+1}$ are *equivalent* if there exists an integer n such that $a = \widetilde{\varphi}^n b$. We denote by W^{n+1} the set of equivalence classes, and by $\pi: \widetilde{W}^{n+1} \longrightarrow W^{n+1}$ the natural projection onto this set. It is easily seen that there exists a unique smooth manifold structure on W^{n+1} such that π becomes a local diffeomorphism. From the fact that $\widetilde{\varphi}$ commutes with $\widetilde{\varphi}_t$, it follows that $\widetilde{\varphi}_t$ preserves the equivalence relation on $\widetilde{W}^{n+1}$ and consequently induces a flow on W^{n+1} (which we denote by φ_t). Finally, it is easily seen that the equivalence classes of points of the form $x \times 0$ $(x \in M^n)$

provide a smooth manifold in W^{n+1}, diffeomorphic to M^n, which constitutes a global transversal for the flow φ_t, and that the succession diffeomorphism on this manifold corresponds to φ.

EXERCISE. Construct the suspension for the diffeomorphism of Example 6b).

DEFINITION. We say that the diffeomorphism $\varphi: M \longrightarrow M$ is *contained in a flow*, if there exists a one-parameter group of homeomorphisms φ_t of M such that $\varphi = \varphi_1$. If φ_t is a group of diffeomorphisms depending differentiably upon t, then we shall say that φ is *contained in a smooth flow*.

PROBLEM. Prove that, for any manifold M, there exists a diffeomorphism φ: $M \longrightarrow M$ which is not contained in a flow.

Hint. Examine the set of periodic points of a diffeomorphism which is included in a flow.

§3. Spaces of dynamical systems. Equivalence relations

Let M be a compact C^∞-manifold. We denote by $\text{Diff}^r(M)$ $(1 \leqslant r \leqslant \infty)$ the set of C^r-diffeomorphisms of M. For any k, $1 \leqslant k \leqslant r \leqslant \infty$, we may convert $\text{Diff}^r(M)$ into a topological (function) space, introducing the topology of C^k-convergence (i.e. uniform convergence together with that of all derivatives of order up to k) in $\text{Diff}^r(M)$. We denote the topological space thus obtained by $\text{Diff}^r_k(M)$. The space Diff^r_k is a complete metric space and possesses the structure of a Banach C^∞-manifold (i.e., one may choose a neighborhood of each point which is homeomorphic to an open ball in a fixed Banach space B such that the transition functions will be C^∞ on B).

Let E be some equivalence relation on Diff^r.

1) For example, let us say that $f, g \in \text{Diff}^r(M)$ are equivalent if there exists a C^l-diffeomorphism $h: M \longrightarrow M$ $(l \leqslant r)$ such that the following diagram is commutative:

$$\begin{array}{ccc} M & \xrightarrow{\ f\ } & M \\ {\scriptstyle h}\big\downarrow & & \big\downarrow{\scriptstyle h} \\ M & \xrightarrow{\ g\ } & M \end{array}$$

This equivalence relation is called *C^l-conjugacy*.

2) If we merely require that the mapping h in the previous diagram be a homeomorphism, then we obtain the relation of topological conjugacy.

3) If we impose further conditions on the homeomorphism h, we obtain various other equivalence relations. For example, we may require that h be homotopic to the identity mapping of M.

4) Below, we shall encounter still another interesting equivalence relation, known as *Ω-conjugacy*.

To every equivalence relation E on the set $\text{Diff}^r(M)$, there corresponds a certain notion of stability on $\text{Diff}^r_k(M)$.

DEFINITION. Let E be an equivalence relation on $\text{Diff}^r(M)$. The diffeomorphism $f \in \text{Diff}^r_k(M)$ is said to be *E-stable* if there exists a neighborhood $U = U(f)$ of the point

f in the space $\mathrm{Diff}^r_k(M)$ such that any diffeomorphism $g \in U(f)$ is E-equivalent to f: $g \underset{E}{\sim} f$.

We emphasize that the notion of E-stability in general depends upon the choice of the topology on Diff^r.

In the case where E is the relation of topological equivalence, a stable diffeomorphism is said to be *structurally stable* or *rough.*

The notion of structural stability or roughness, which has turned out to be especially fruitful, first appeared in 1937, in a paper of Andronov and Pontrjagin, in connection with flows [1] (i.e., dynamical systems with continuous time). The definitions of topological equivalence and structural stability for a flow are not entirely parallel to the corresponding definitions for a diffeomorphism, and will be given below.

For a compact manifold M, the space of smooth flows on M may be identified with the linear space $\mathfrak{X}^r(M)$ of C^r-smooth vector fields on M. Provided with a C^r-norm, $\mathfrak{X}^r$ becomes a Banach space. Given a vector field X on M, for an arbitrary homeomorphism h the objects $X \circ h$ and $h \circ X$ are in general not defined, and we must pass from the field X to the one-parameter group of diffeomorphisms φ_t generated by X. Then, for any fixed t, the homeomorphisms $\varphi_t \circ h$ and $h \circ \varphi_t$ are well-defined. However, the definition of topological conjugacy which suggests itself (that is, $\{\varphi_t\} \sim \{\psi_t\} \iff \exists h \forall t\; h \circ \psi_t = \varphi_t \circ h$) turns out to be unsuccessful, and must be modified. For otherwise, e.g., the periods of closed trajectories turn out to be invariants, and flows having identically disposed trajectories, and differing only in the velocity of motion along these trajectories, would not be equivalent.

DEFINITION. Two smooth flows $\{\varphi_t\}$ and $\{\psi_t\}$ on the manifold M are said to be *topologically equivalent* if there exists a homeomorphism $h: M \longrightarrow M$ which transforms the trajectories of $\{\varphi_t\}$ into the trajectories of $\{\psi_t\}$ with preservation of orientation.

DEFINITION. A flow $\varphi_t \in \mathfrak{X}^r$ is said to be *structurally stable* if there exists a neighborhood U of φ_t in $\mathfrak{X}^r(M)$ such that any flow $\psi_t \in U$ is topologically equivalent to φ_t.

DEFINITION. Two smooth flows $\{\varphi_t\}$ and $\{\psi_t\}$ in $\mathfrak{X}^r(M)$ will be called *C^l-conjugate* if there exists a diffeomorphism $h \in \mathrm{Diff}^l(M)$ such that, for any $t \in \mathbf{R}$, $\psi_t = h^{-1} \circ \varphi_t \circ h$.

EXERCISE. Show that the dynamical systems of Examples 1 and 2 are not rough. Show that the systems of Examples 3–5 lack stability with respect to C^1-conjugacy.

§4. Classification of dynamical systems

First of all, it should be observed that there is no hope of classifying all dynamical systems without exception. There are many reasons for this. For example, the structure of the set of fixed points must certainly be an invariant of the dynamical system for any reasonable equivalence relation; however, any closed subset of a manifold M is the set of fixed points of some C^∞-smooth dynamical system. But we are inclined to regard the problem of describing all closed subsets of a manifold M as unsolvable.

However, it will be shown below that, for all systems belonging to a certain open set which is everywhere dense in $\mathrm{Diff}_r^r(M)$ or in $\mathfrak{X}^r(M)$, the set of fixed points is finite. Thus, we are led to the necessity of excluding from consideration a closed nowhere dense set in the space of dynamical systems, consisting of systems for which the set of fixed points is infinite (which we regard as an exceptional, nontypical situation). Some other nontypical situations (also encountered only upon a closed, nowhere dense set) will be described below.

The second remark which should be made is that, for the global study of a dynamical system (on the entire manifold M), C^k-conjugacy (even for $k = 1$) is too fine an equivalence relation. One of the reasons for this is that, if a dynamical system has a fixed point, then the eigenvalues of the linear portion of the dynamical system at this point (the rigorous definition will be given below) are C^1-invariants; such invariants, taken with respect to all fixed and periodic points, do not, in general, constitute a complete set of invariants.*

PROBLEM. Consider the flow φ of Example 3. Show that there are flows, arbitrarily close to φ in the C^∞-topology, which coincide with φ on the sets $|z| < 1$ and $|z| > 2$, and are not C^1-equivalent to φ.

We shall now state a theorem, which is basically due (like the formulation of the notion of roughness) to Andronov and Pontrjagin.

THEOREM (ANDRONOV-PONTRJAGIN AND DE BAGGIS-PEIXOTO). *The set of structurally stable flows is everywhere dense (and open) in $\mathfrak{X}^r(M^2)$ for any compact two-dimensional manifold M^2.*

Andronov and Pontrjagin studied flows on the two-dimensional sphere and disk, and gave a number of simple geometrical conditions, necessary and sufficient for structural stability. Peixoto [2], [3] proved that the same conditions are necessary and sufficient on any compact manifold and that these conditions characterize an open everywhere dense set. The precise formulation of the Andronov-Pontrjagin conditions will be given below.

As has become clear during the last ten years, the analogue of the Andronov-Pontrjagin theorem, unfortunately, does not hold for manifolds of dimension greater than two. On the one hand, this is manifested in the fact that, in attempting to generalize the geometrical conditions for roughness, we obtain only sufficient conditions, since there exist rough systems with very complicated structures (which will be discussed below). On the other hand, we have the following result:

THEOREM OF SMALE. *For any manifold M^m with $m \geqslant 2$, there exists an open set in* $\mathrm{Diff}_r^r(M)$ $(1 \leqslant r \leqslant \infty)$ *consisting of nonrough diffeomorphisms, and for $m \geqslant 3$ there exists an open set of nonrough flows in $\mathfrak{X}^r(M)$.*

**Translator's note.* The purport of this phrase is obscure. Perhaps the intended meaning is "satisfactory" or "meaningful" rather than "complete".

Smale [4] constructed an example of an open set of nonrough diffeomorphisms in $\mathrm{Diff}(T^3)$, where T^3 denotes the three-dimensional torus. Application of the suspension provided an analogous example for flows on a certain four-dimensional manifold.

The construction of the same examples on an arbitrary manifold (the greatest differences arise for diffeomorphisms on two-dimensional (see [5]) and flows on three-dimensional manifolds) was subsequently carried out by a number of authors.

Concerning the formulation of the theorem of Smale, it is worth making one remark. One might attempt to save the situation by introducing a finite number of invariants. In the problem of classifying matrices up to similarity, for example, stability is also lacking: in the neighborhood of any matrix one can find a matrix which is not similar to it. However, on an open, everywhere dense set of matrices there exist a finite number of invariants (the eigenvalues), which completely resolve the classification problem on this set. But a slight modification of the construction of Smale allows one to construct an open set of diffeomorphisms on which the number of invariants of topological conjugacy is infinite.

§5. Fixed points and periodic trajectories

1. Let $f: M^n \longrightarrow M^n$, $fx_0 = x_0$. In local coordinates with center at x_0, one may write $fx = Ax + \cdots$. Writing an analogous expansion in another system of local coordinates, we obtain, in place of A, a matrix similar to A. The appropriately defined linear operator $f'|_{x_0}$ on the tangent space $TM^n|_{x_0}$, the matrix of which, in the basis $x_1, \ldots, x_n$ (or, more precisely, in the basis $\partial/\partial x_1|_{x_0}, \ldots, \partial/\partial x_n|_{x_0}$) is A, is called the "linear part" of the diffeomorphism f at the fixed point x_0. The eigenvalues of $f'|_{x_0}$ are called the *eigenvalues of the diffeomorphism f at the fixed point x_0*. If the point x_0 is periodic, with period k, then the preceding construction can be applied to f^k. The eigenvalues of f^k at x_0 are called the *eigenvalues of the diffeomorphism f at the periodic point x_0*.

If the vector field X vanishes at the point x_0, then the corresponding system of differential equations, in the local coordinates $x_1, \ldots, x_n$, has the form $\dot{x} = Ax + \cdots$. Again, the matrix A constitutes the matrix of a properly defined linear operator X' on $TM^m|_{x_0}$, and the eigenvalues of A are known as the *eigenvalues of the vector field at the stationary point x_0*.

Finally, if the vector field X has a periodic trajectory γ, then we apply the preceding definition to the succession function, and we shall refer to the eigenvalues of the succession function as the *eigenvalues of the periodic trajectory γ*.

EXERCISE. Let φ_t be a flow corresponding to the vector field X, let T be the period of γ, let $\lambda_1, \ldots, \lambda_{n-1}$ be the eigenvalues of γ, and let $x_0 \in \gamma$. Then 1, $\lambda_1, \ldots, \lambda_{n-1}$ are the eigenvalues of φ_T at the fixed point x_0.

DEFINITION. A nonsingular linear operator A (in a finite-dimensional space) is said to be *hyperbolic* if its spectrum does not intersect the circle $|\lambda| = 1$.

Correspondingly, a fixed or periodic point of a diffeomorphism, or a periodic trajectory of a flow, is said to be hyperbolic if none of its eigenvalues has modulus 1.

A fixed point x_0 of a flow φ_t generated by a vector field X is said to be hyperbolic if x_0 is a hyperbolic fixed point of the diffeomorphism φ_1. This condition is satisfied if and only if the real parts of the eigenvalues of X at x_0 are different from zero.

2. We consider the important special case where all the eigenvalues of the diffeomorphism φ at the fixed point x_0 are of modulus less than 1. In this case, the local structure of φ in the neighborhood of x_0, from the topological point of view, is very simple:

THEOREM 1. *If the abovementioned diffeomorphism φ preserves orientation, then, locally, φ is topologically conjugate to the diffeomorphism $\psi\colon x \longrightarrow \frac{1}{2}x$, $x \in \mathbf{R}^n$. If φ reverses orientation, then, locally, φ is topologically equivalent to the diffeomorphism $\psi \circ \sigma$, where σ denotes the reflection with respect to the hyperplane $x_1 = 0$.*

We note that the statement of this theorem is only concerned with topological (and definitely not with differentiable) conjugacy.

An equivalent formulation of Theorem 1 is as follows:

THEOREM 1′. *Suppose that the eigenvalues of the diffeomorphisms $\varphi_i\colon \mathbf{R}_i^n \longrightarrow \mathbf{R}_i^n$ $(i = 1, 2)$ at the fixed point 0 have moduli less than 1. If φ_1 and φ_2 both preserve (or both reverse) the orientation of $\mathbf{R}^n$, then φ_1 and φ_2 are locally topologically conjugate; that is, there exists a local homeomorphism h such that $\varphi_2 \circ h = h \circ \varphi_1$.*

Still another equivalent formulation is useful:

THEOREM 1″. 1) *If the eigenvalues of the diffeomorphism $\varphi\colon \mathbf{R}^n \longrightarrow \mathbf{R}^n$ at the fixed point 0 are less than one in modulus, then φ is locally topologically conjugate to its own linear part.*

2) *Any two linear diffeomorphisms of $\mathbf{R}^n$ with eigenvalues less than one in modulus are topologically conjugate if they lie in the same connected component of the group $GL(n, \mathbf{R})$.*

An analogous theorem is valid for flows:

THEOREM 2. *Suppose that the linear part of a vector field X at the stationary point $0 \in \mathbf{R}^n$ has eigenvalues with negative real parts. Then there exists a homeomorphism h, defined in some neighborhood of the origin, such that the equation $(h \circ e^{-t})x = \varphi_t \circ h$ holds whenever both sides are defined.**

One may likewise formulate Theorems 2′ and 2″, analogous to Theorems 1′ and 1″.

PROOF OF THEOREM 2. By a theorem of Ljapunov, there exists a neighborhood D of 0 such that $\bigcap_{t \geqslant 0} \varphi_t D = 0$. Moreover, one may assume that D is the interior of some ellipsoid, and consequently that there exists a diffeomorphism f of the standard sphere $x_1^2 + \cdots + x_n^2 = 1$ onto $\partial\overline{D}$ (as f one may take, for example, the projection

*Translator's note. φ_t, of course, denotes the flow generated by X.

along the rays emanating from the origin). We define the continuous mapping h in the following manner. Let $y = e^{-t}x$, where $|x| = 1$ and $t > 0$. We set $h(y) = \varphi_t \circ f(x)$ and $h(0) = 0$. It is easily seen that the mapping so obtained is a homeomorphism of the set $x_1^2 + \cdots + x_n^2 \leqslant 1$ onto D and satisfies the condition $\forall t, h \circ e^{-t} = \varphi_t \circ h$. An analogous construction for diffeomorphisms turns out to be slightly more complicated.

We shall now sketch the proof of Theorem 1′. It is useful to visualize the entire construction in two- or three-dimensional space. By a theorem of Ljapunov, there exists a region $D_i \supset 0$ in $\mathbf{R}_i^n$, with boundary $\partial \overline{D}_i = S_i^{n-1}$, all the points of which approach 0 under the action of φ_i^n as $n \longrightarrow +\infty$. Moreover, as D_i one may take the interior of a sufficiently small ellipsoid. We construct the conjugating homeomorphism h in several steps. Let f be any diffeomorphism of the sphere S_1^{n-1} onto S_2^{n-1}. We set $h|_{S_1^{n-1}} = f$. Hence h is defined on $\varphi_1 S_1^{n-1}$:

$$h\big|_{\varphi_1 S_1^{n-1}} = \varphi_2 \circ h\big|_{S_1^{n-1}} \circ \varphi_1^{-1}.$$

Thus the homeomorphism h has already been defined on the set $S_1^{n-1} \cup \varphi_1 S_1^{n-1}$. This set forms the boundary of the "annulus" (or "spherical shell") $\overline{M}_1$, where $M_1 = D_1 \backslash \varphi_1 D_1$. We note that the set M_1 does not intersect its own translates; that is, $\forall k \neq 0$, $\varphi_1^k M_1 \cap M_1 = \varnothing$, and the closure of M_1 intersects its translate only in a sphere: $\overline{M}_1 \cap \varphi_1 \overline{M}_1 = \varphi_1 S_1^{n-1}$. Finally, since all points of the ball D_1 approach 0 under the action of φ_1^n as $n \longrightarrow +\infty$, we have $\bigcup_{k \geqslant 0} \varphi_1^k M_1 = D_1 \backslash 0$. For the construction of h on M_1, we require the following fact.

LEMMA. *If φ_1 and φ_2 either both preserve orientation, or both reverse orientation, then the homeomorphism h: $\partial(\overline{M}_1) \longrightarrow \partial(\overline{M}_2)$ may be extended to a homeomorphism h: $\overline{M}_1 \longrightarrow \overline{M}_2$. In other words, the homeomorphism of the boundary of the "spherical shell" can be extended to a homeomorphism of the entire spherical shell.*

By this lemma, we may assume that the homeomorphism h has been constructed on $\overline{M}_1$. The relation $\varphi_2 \circ h = h \circ \varphi_1$, which we wish to obtain, implies that on $\varphi_1^k M_1$ we must define h in the following manner:

$$h\big|_{\varphi_1^k M_1} = \varphi_2^k \circ h\big|_{M_1} \circ \varphi_1^{-k}.$$

As is easily seen, the formulas thus obtained define a homeomorphism h: $D_1 \backslash 0 \longrightarrow D_2 \backslash 0$, which, by virtue of what has been stated above, may be extended to a homeomorphism h: $D_1 \longrightarrow D_2$ by setting $h(0) = 0$. Finally, by the construction itself, h satisfies the relation $\varphi_2 \circ h = h \circ \varphi_1$.

The proof of the lemma is simple enough and may be carried out in various ways. We shall not dwell upon this, and merely remark that no topological problems (of types such as the "problem of the annulus," "exotic spheres," or the connectivity of the group of homeomorphisms of the sphere) can arise here, since one may assume that the diffeomorphisms φ_1 and φ_2 are arbitrarily close to linear diffeomorphisms, and that the manifolds $\varphi_i S_i^{n-1}$ are arbitrarily close to ellipsoids.

Hint. In the proof of the lemma, the following facts may be helpful.

a) $\overline{M}_i$ is diffeomorphic to $S^{n-1} \times [0, 1]$.

b) If f is a diffeomorphism of the sphere $|x| = 1$ $(x = (x_1, \ldots, x_n))$, close to a diffeomorphism defined by the formula $x \longrightarrow Ax/|Ax|$, $|x| = 1$, where $\det A > 0$, then there exists a family of diffeomorphisms $g_t: S^{n-1} \longrightarrow S^{n-1}$, depending continuously upon t, such that g_0 is the identity diffeomorphism and $g_1 = f$.

We recall the proof of Theorem 2 and consider the open set $W^s = \bigcup_{-\infty < t \leqslant 0} \varphi_t D$. Clearly, W^s is invariant with respect to φ_t, and any point of W^s approaches 0 under the action of φ_t as $t \longrightarrow +\infty$. It is easy to show that W^s is homeomorphic (and even diffeomorphic) to $\mathbf{R}^n$, and that the action of φ_t on W^s is topologically conjugate to that of the one-parameter group of dilations ψ_t of $\mathbf{R}^n$: $\psi_t(x) = e^{-t}x$.

Similarly, examining the proof of Theorem 1′, one can deduce that the set $W^s = \bigcup_{-\infty < n \leqslant 0} \varphi_1 D_1$ is homeomorphic (and even diffeomorphic) to $\mathbf{R}^n$, and that the diffeomorphism $\varphi_1|_{W^s}$ is topologically conjugate to either the diffeomorphism $\psi: \mathbf{R}^n \longrightarrow \mathbf{R}^n$, $\psi(x) = x/2$, or the diffeomorphism $\sigma \circ \psi$, where σ is the reflection in the hyperplane $x_1 = 0$.

In both of the cases considered above, the set W^s (s stands for *stable*) consists of those and only those points which approach the stationary point as $t \longrightarrow +\infty$ (or $n \longrightarrow +\infty$).

DEFINITION. A nonsingular linear operator A: $\mathbf{R}^n \longrightarrow \mathbf{R}^n$ is said to be *contractive* if all the eigenvalues of A are of modulus less than 1, and *expansive* if all the eigenvalues of A are of modulus greater than 1, i.e. if A^{-1} is contractive.

A fixed point x_0 of a diffeomorphism φ is said to be *attractive* (*repulsive*) if the linear part of φ at x_0 is a contractive (expansive) linear operator.

A periodic point x_0 of a diffeomorphism φ is said to be *attractive* (*repulsive*) if x_0 is an attractive (repulsive) point of the diffeomorphism φ^k (where k is the period of x_0).

A fixed point x_0 of the flow φ_t is said to be *attractive* (*repulsive*) if x_0 is an attractive (repulsive) point of the diffeomorphism φ_1.

Finally, an *attractive* (*repulsive*) *periodic trajectory* of the flow φ_t is defined by means of the succession function.

The theorems proved above, concerning the topological structure of diffeomorphisms and flows in the neighborhood of attractive fixed points, carry over, with the appropriate modifications in formulation, to repulsive fixed points as well as attractive or repulsive periodic points and trajectories.

By slightly modifying the proof of Theorem 1′, one can obtain a proof of the following fact.

ASSERTION. *Let $x_0 \in M^n$ be an attractive fixed point of the diffeomorphism φ, let $W^s(x_0)$ be the set of all points of M^n which are attracted to x_0 under the action of the positive powers of φ, and let U be an open set in $W^s(x_0)$ satisfying the following conditions*:

1) $x_0 \in U$.
2) ∂U *is diffeomorphic to* S^{n-1}.
3) $\overline{\varphi U} \subset U$.

Then, for any diffeomorphism $\widetilde{\varphi}$ *which is sufficiently close to* φ *in the* C^1*-topology, there exists a homeomorphism* $h\colon \overline{U} \longrightarrow \overline{U}$ *such that* $h \circ \varphi = \widetilde{\varphi} \circ h$. *Moreover, one may assume that h coincides with the identity in some neighborhood of* $\partial \overline{U}$. *An analogous statement also holds for flows.*

REMARK. Thus we see that, for a C^1-small variation of φ, the structure of φ, from a topological viewpoint, does not change (in any case, in a small neighborhood of x_0). The set $W^s(x_0)$, however, may change strongly for an arbitrarily small variation of φ (although, from the above assertion, one can easily deduce that this set cannot diminish too strongly).

EXAMPLE. Consider the following system, on the (x, y)-plane, depending on the parameter ϵ:

$$\begin{pmatrix} \dot{x} \\ \dot{y} \end{pmatrix} = \begin{pmatrix} y \\ -x \end{pmatrix} - \begin{pmatrix} x \\ y \end{pmatrix} \cdot [(x^2 + y^2 - 1)^2 + \epsilon].$$

It is easily seen that, for any ϵ, the origin is an attractive fixed point (i.e. a stable focus).

For $\epsilon = 0$, the system has the limit cycle $x^2 + y^2 = 1$, and in this case $W^s(0) = \{(x, y)\colon x^2 + y^2 < 1\}$, while for any $\epsilon > 0$ there are no limit cycles, and $W^s(0) = \mathbf{R}^2$.

REMARK. The boundary $\partial(W^s)$ of the set of points which are attracted to x_0 is clearly a closed set which is invariant with respect to φ (or φ_t). This set may have a very complicated structure (see the lectures of A. B. Katok contained in this volume).

EXERCISE. Using the above assertion, show that the systems of Examples 3 and 4 are structurally stable with respect to the C^1-topology on $\mathrm{Diff}^1(S^2)$ or $\mathfrak{X}^1(S^2)$.

3. We now pass to the discussion of the general hyperbolic case, where some of the eigenvalues lie inside the unit circle, and some outside. We begin with the study of a linear operator. The following lemma is obvious.

LEMMA. *Let* $A\colon \mathbf{R}^n \longrightarrow \mathbf{R}^n$ *be a hyperbolic linear operator. Then there exists a decomposition of* $\mathbf{R}^n$ *into a direct sum* $\mathbf{R}^n = \mathbf{R}^{k_1} \oplus \mathbf{R}^{k_2}$ *of two subspaces, invariant with respect to A, such that* $A|_{\mathbf{R}^{k_1}}$ *is a contractive linear operator and* $A|_{\mathbf{R}^{k_2}}$ *is an expansive linear operator. Thus, a linear hyperbolic diffeomorphism is the direct product of a contractive and an expansive linear diffeomorphism. The set of points which approach the fixed point as* $n \longrightarrow +\infty$ *is a smooth manifold* $(\mathbf{R}^{k_1})$ *passing through the fixed point. Similarly, the set of points which approach the fixed point as* $n \longrightarrow -\infty$ *is a smooth manifold* $(\mathbf{R}^{k_2})$ *passing through the fixed point. Every point* $x \notin \mathbf{R}^{k_1} \cup \mathbf{R}^{k_2}$ *leaves any small neighborhood of the fixed point both for* $n \longrightarrow +\infty$ *and for* $n \longrightarrow -\infty$.

We see at once that an arbitrary hyperbolic fixed point of a nonlinear diffeomorphism possesses, locally, all the properties which have been enumerated for a linear hyperbolic fixed point.

THEOREM OF HADAMARD AND PERRON. *Let x_0 be a hyperbolic fixed point of a diffeomorphism $\varphi: M \longrightarrow M$. There exists a neighborhood U of x_0 such that the following conditions are satisfied*:

a) *The set of points of U which approach x_0 as $n \longrightarrow +\infty$ and do not leave U is a smooth manifold W^s passing through x_0.*

b) *The set of points of U which approach x_0 as $n \longrightarrow -\infty$ and do not leave U is a smooth manifold W^u (u stands for* unstable) *passing through x_0.*

Furthermore, 1) $x_0 \in W^s$ *is an attractive fixed point of* $\varphi|_{W^s}$ *and* 2) $x_0 \in W^u$ *is a repulsive fixed point of* $\varphi|_{W^u}$.

3) *The tangent spaces of W^s and W^u are invariant with respect to the linear part of φ and provide a decomposition of $\varphi'|_{x_0}$ into the direct sum of a contractive and an expansive operator.*

4) *The smoothness of W^s and W^u is not less than the smoothness of φ.*

THEOREM OF GROBMAN AND HARTMAN [6], [7]. *Let x_0 be a hyperbolic fixed point of the diffeomorphism $\varphi: M \longrightarrow M$. Then, in some neighborhood of x_0, φ is topologically conjugate to its own linear part; that is, there exists a homeomorphism $h: U \longrightarrow \varphi(U) \subset \mathbf{R}^n$ such that the following diagram is commutative:*

$$\begin{array}{ccc} M & \xrightarrow{\ \varphi\ } & M \\ {\scriptstyle h}\downarrow & & \downarrow{\scriptstyle h} \\ \mathbf{R}^n & \xrightarrow{\ \varphi'|_{x_0}\ } & \mathbf{R}^n \end{array}$$

More precisely, $(h \circ \varphi)(x) = (\varphi'|_{x_0} \circ h)(x)$ *whenever both the left and right sides are defined.*

Since every linear hyperbolic diffeomorphism is the direct product of a contractive linear diffeomorphism on a space of dimension k_1 and an expansive linear diffeomorphism on a space of dimension k_2, where $0 \leqslant k_1, k_2 \leqslant n$ and $k_1 + k_2 = n$, the Grobman-Hartman theorem and (e.g.) Theorem 1 imply the following assertion:

THEOREM. *Let x_0 be a hyperbolic fixed point of the diffeomorphism $\varphi: M^n \longrightarrow M^n$, let* $k_1 = \dim W^s(x_0)$ *and* $k_2 = \dim W^u(x_0)$, *and set*

$$\sigma_1 = \begin{cases} +1 \text{ if } \varphi|_{W^s} \text{ preserves orientation, or if } k_1 = 0, \\ -1 \text{ otherwise;} \end{cases}$$

$$\sigma_2 = \begin{cases} +1 \text{ if } \varphi|_{W^u} \text{ preserves orientation, or if } k_2 = 0, \\ -1 \text{ otherwise.} \end{cases}$$

Then the numbers k_1, k_2, σ_1 and σ_2 constitute a complete set of invariants for local topological conjugacy; in particular, every hyperbolic fixed point is locally structurally stable.

EXERCISE. Formulate the Hadamard-Perron theorem and the Grobman-Hartman theorem for diffeomorphisms of a Banach space.

The three theorems stated above were first formulated for flows, and in that case the last of these assumes a somewhat simpler form, since all orientations are preserved in the case of a flow.

It is not difficult to formulate analogous theorems for periodic points and periodic solutions.

4. In order to deal with global rather than local objects, it is often convenient to perform the following construction. Let $\varphi: M^n \longrightarrow M^n$ be a diffeomorphism and let x_0 be a hyperbolic fixed point of φ. Then, in some system of local coordinates in the neighborhood of x_0, φ has the form $\varphi(x) = Ax + f(x)$, where $f(0) = 0$ and $(\partial f/\partial x)(0) = 0$. Choosing a C^∞ truncating function ρ satisfying the condition

$$\rho(t) = \begin{cases} 1, & |t| < \epsilon^2, \\ 0, & |t| > \epsilon, \end{cases}$$

for a sufficiently small ϵ, we may consider the diffeomorphism $\widetilde{\varphi}(x) = Ax + f(x)\rho(|x|)$, defined on the entire space $\mathbf{R}^n$ and arbitrarily close in the C^1 metric to the linear diffeomorphism $x \longrightarrow Ax$. Locally, the diffeomorphisms φ and $\widetilde{\varphi}$ are diffeomorphic (i.e., differentiably equivalent). Moreover, the Hadamard-Perron and Grobman-Hartman theorems turn out to be globally valid for $\widetilde{\varphi}$.

THEOREM. *Let $\widetilde{\varphi}$ be a linear hyperbolic diffeomorphism of $\mathbf{R}^n$. Then there exists an $\epsilon = \epsilon(\widetilde{\varphi})$ such that, for any diffeomorphism $\varphi: \mathbf{R}^n \longrightarrow \mathbf{R}^n$ satisfying the condition*

$$\forall x, \quad |\widetilde{\varphi}(x) - \varphi(x)| < \epsilon \quad \text{and} \quad \|\widetilde{\varphi}'|_x - \varphi'|_x\| < \epsilon,$$

the following statements hold:

1) *There exists a homeomorphism $h: \mathbf{R}^n \longrightarrow \mathbf{R}^n$, close to the identity, such that $\varphi \circ h = h \circ \widetilde{\varphi}$.*

2) *The invariant manifolds of the unique fixed point x_0 of φ have the same smoothness as φ.*

The rigorous proof of assertion 1 of this theorem, and, consequently, the proof of the Grobman-Hartman theorem, will be given below.

REMARKS. 1) The Grobman-Hartman theorem implies the existence of topological (that is, merely continuous) invariant manifolds passing through a hyperbolic fixed point (these manifolds correspond to the invariant manifolds of a linear transformation).

2) One way of proving the Grobman-Hartman theorem consists in the construction of a conjugating homeomorphsim h, first on the invariant manifolds, the existence of which is guaranteed by the Hadamard-Perron theorem, and then on their complement. The latter part of the construction is facilitated by the fact that the trajectory of any point $x \notin W^s(x_0) \cup W^u(x_0)$ intersects a sufficiently small neighborhood of x_0 only in a finite number of points.

3) It is interesting to note that the Grobman-Hartman theorem in the finite-dimensional case is a consequence of the Hadamard-Perron theorem for diffeomorphisms of a Banach space.

5. We shall now describe some geometrical concepts which will enable us to prove the Hadamard-Perron and Grobman-Hartman theorems. We confine ourselves to the two-dimensional case.

Let $\widetilde{\varphi}$: $\mathbf{R}^2 \longrightarrow \mathbf{R}^2$ be the linear diffeomorphism of $\mathbf{R}^2$ given by

$$\begin{pmatrix} x \\ y \end{pmatrix} \longrightarrow \begin{pmatrix} 2x \\ \frac{1}{2}y \end{pmatrix}.$$

Obviously, the origin is a hyperbolic fixed point of $\widetilde{\varphi}$. The manifold $W^s(0)$ is the y-axis, and $W^u(0)$ is the x-axis. The family of straight lines parallel to the x-axis is invariant with respect to $\widetilde{\varphi}$; that is, every straight line of this family is transformed by the diffeomorphism $\widetilde{\varphi}$ into some (in general, different) straight line of this family. Only one line of this family, namely the x-axis, is mapped into itself by the transformation $\widetilde{\varphi}$. Two points $a, b \in \mathbf{R}^2$ belong to one straight line of the family under consideration if and only if $\rho(\widetilde{\varphi}^n a, \widetilde{\varphi}^n b) \longrightarrow 0$ as $n \longrightarrow -\infty$. In particular, the restriction of $\widetilde{\varphi}$ to $W^u(0)$ is an expansive diffeomorphism. The distance between any two lines γ_1 and γ_2 of the family under consideration, in the metric of C,* decreases exponentially as $n \longrightarrow +\infty$:

$$\rho(\widetilde{\varphi}^n \gamma_1, \widetilde{\varphi}^n \gamma_2) \xrightarrow[n \to +\infty]{} 0.$$

Finally, the lines of this family are in one-to-one correspondence with the points of the manifold $W^s(0)$.

Analogous properties are likewise valid for the family of lines parallel to the y-axis.

It turns out that, for any C^1-small perturbation φ of the diffeomorphism $\widetilde{\varphi}$, there exist two families of curves, close to the families of lines considered above, and possessing analogous properties.

For the proof of existence of such families, we consider the behavior of the tangent vectors of $\mathbf{R}^2$ under the diffeomorphisms φ and $\widetilde{\varphi}$. The Jacobian matrix of the transformation $\widetilde{\varphi}$ at any point $a \in \mathbf{R}^2$ is

$$\begin{pmatrix} 2 & 0 \\ 0 & \frac{1}{2} \end{pmatrix} = A.$$

Clearly, for any diffeomorphism φ which is sufficiently close to $\widetilde{\varphi}$ (closeness being understood in the C^1-sense) on the whole of $\mathbf{R}^2$, the Jacobian matrix of φ at any point $a \in \mathbf{R}^2$ differs from the matrix A by at most ϵ.

Hence, one may deduce that for sufficiently small ϵ the following assertions are valid:

1) *Under the application of* $\varphi'|_a$, *the length of any vertical tangent vector with initial point* $a \in \mathbf{R}^2$ *decreases by a factor of at least* $2/3$.

**Translator's note.* The rendering "metric C" is also possible. In either case the identity of the metric referred to is obscure; however, the statement would obviously be valid for any reasonably defined metric.

2) *Under the application of $\varphi'|_a$, any tangent vector e which deviates from the horizontal direction by less than 45° (in absolute value) is again transformed into a tangent vector, the length of which is not less than 4/3 the length of e, and forming an angle of less than 45° (in absolute value) with the horizontal direction.*

Moreover, we have the following statements:

3) *There exists a constant $\lambda < 1$ such that, for any two tangent vectors e_1, e_2 with origin at the point a,*

$$\measuredangle(\varphi'|_a e_1, \varphi'|_a e_2) \leqslant \lambda \measuredangle(e_1, e_2),$$

provided that both e_1 and e_2 form an angle of less than 45° (in absolute value) with the horizontal direction.

4) *Assertions* 1)–3) *remain valid when φ is replaced by φ^{-1} and the word "horizontal" is replaced by "vertical" (and vice versa).*

We are now in a position to prove the Hadamard-Perron theorem for the diffeomorphism φ. We consider the set D of curves of the form $y = f(x)$, where the function f is differentiable and $\forall x, |f'(x)| < 1$. It follows from 2) that, under the action of φ, any curve of D is transformed into a curve of D. From 1) it is easy to deduce that the mapping $\hat{\varphi}: D \to D$ is contractive in the metric of C^0,* and consequently has a unique fixed point γ_+, which is the graph of a function $y = f_+(x)$. Since γ_+ may be determined by successive approximations, the function $f_+(x)$ is the limit of a uniformly convergent sequence of differentiable functions $y = f_n(x)$ satisfying the condition $|f_n'(x)| \leqslant 1$. To prove the differentiability of the curve γ_+ (that is, the differentiability of $f_+(x)$) we observe that along with assertions 1)–4) one may also prove the following assertion:

5) *If the diffeomorphism φ belongs to the class C^k and the derivatives of φ up to order k (inclusive) are uniformly bounded on $\mathbf{R}^2$, then there exist constants $C_1 = 1$, $C_2, \ldots, C_k$ such that φ transforms the graph of a function $y = f(x)$, where $|f^{(i)}(x)| < C_i$, $i = 1, \ldots, k$, into the graph of a function $y = g(x)$, where $|g^{(i)}(x)| < C_i$, $i = 1, \ldots, k$.*

This assertion is obvious if φ is close to $\tilde{\varphi}$ in the C^k-topology, and may be proved in a straightforward manner if φ is close to $\tilde{\varphi}$ only in the C^1-topology.

We now assume that $f_1(x)$ satisfies the conditions $|f^{(i)}(x)| \leqslant C_i$, $i = 1, \ldots k$. Then the functions $f_n(x)$ satisfy the same conditions, and by Arzelà's lemma one may select a subsequence of the sequence $f_n(x)$ which converges to a $(k-1)$-times continuously differentiable function, and therefore the invariant curve γ_+ belongs to the class C^{k-1}, where k is the smoothness of φ. The proof that γ_+ belongs, in fact, to the class C^k requires consideration of the modulus of continuity of the kth derivative of the elements of the sequence $f_n(x)$.

We observe that, at any one of its points, the invariant curve γ_+ which we have obtained forms an angle of less than 45° (in absolute value) with the horizontal direction (in fact, the angle is less in absolute value than some $d(\epsilon)$, where $d(\epsilon) \to 0$ as $\epsilon \to 0$).

*See the previous translator's note.

Applying a similar argument to φ^{-1}, we obtain an invariant curve γ_- which forms an angle of less than 45° (in absolute value) with the vertical direction. The curves γ_+ and γ_- have a unique point of intersection x_0, which is the unique fixed point of φ.

It follows from 2) that the diffeomorphism $\varphi|_{\gamma_+}$ is expansive, and that $\varphi|_{\gamma_-}$ is contractive.

Thus, part of the assertions of the Hadamard-Perron theorem have been proved. It remains to prove that any point $x \notin \gamma_+ \cup \gamma_-$ leaves any neighborhood of x_0. Although this assertion is easy to prove directly, we remark that it follows from the Grobman-Hartman theorem, which will be proved below.

We now proceed to the construction of the required families of curves that are invariant with respect to φ.

We first consider the set P_+ of continuous fields of directions on $\mathbf{R}^2$ which differ from the horizontal direction by not more than 45° at every point of $\mathbf{R}^2$. We introduce a C^0-metric in the set of such fields. It follows from assertion 2) that the set P_+ is invariant with respect to (the mapping induced by) φ; and, moreover, it follows from assertion 3) that φ constitutes a contractive mapping of P_+. The unique fixed point of P_+ under this mapping is a continuous field of directions on $\mathbf{R}^2$ which is invariant with respect to φ. Under the action of $(\varphi')^n$, every tangent vector belonging to this field of directions expands exponentially as $n \longrightarrow +\infty$ and contracts exponentially as $n \longrightarrow -\infty$. The invariant curve γ_+ constitutes an integral curve of the invariant field of directions P_+ which we have constructed.

If P_+ satisfied the conditions of the existence and uniqueness theorem for ordinary differential equations, then, taking the integral curves of P_+, we would obtain a family of curves invariant with respect to φ. Unfortunately P_+ does not satisfy the conditions of any uniqueness theorem, and we must construct the invariant family of curves directly.

We take any smooth family of smooth curves on the plane $\mathbf{R}^2$, C^1-close to the family of all horizontal lines (e.g., the very same family of horizontal lines itself). We denote this family by Γ_0 and consider the families $\Gamma_1, \Gamma_2, \ldots$ obtained from Γ_0 by the actions of $\varphi, \varphi^2, \ldots$, respectively. We fix a point $a \in \mathbf{R}^2$, and consider the curves $l_0, l_1, l_2, \ldots$, belonging to the respective families $\Gamma_0, \Gamma_1, \Gamma_2, \ldots$, which pass through a. Using assertion 1), it is easily shown that $l_0, l_1, l_2, \ldots$ converge uniformly to a limit curve $l_+(a)$, and using the same argument as above, one may show that $l_+(a)$ belongs to the same smoothness class as φ. From the uniqueness of $l_+(a)$ it follows that $l_+(\varphi a) = \varphi l_+(a)$. Similarly, for any point a, one may construct a curve $l_-(a)$. Clearly, $l_+(a)$ is an integral curve of the invariant field of directions P_+, and $l_-(a)$ is an integral curve of P_-. For any two distinct points a_1 and a_2, lying on some curve $l_+(a)$ ($l_-(a)$),

$$\rho(\varphi^n a_1, \varphi^n a_2) \longrightarrow \begin{cases} \infty & \text{as } n \longrightarrow +\infty\ (n \longrightarrow -\infty), \\ 0 & \text{as } n \longrightarrow -\infty\ (n \longrightarrow +\infty). \end{cases}$$

For any two points a_1 and a_2, the curves $l_+(a_1)$ and $l_+(a_2)$ intersect in precisely one point. Assume that the point b does not belong to $l_-(a)$. We show that then $\rho(\varphi^n b, \varphi^n a) \longrightarrow \infty$ as $n \longrightarrow +\infty$. In fact, if we denote by c the point of intersection of $l_+(b)$ and $l_-(a)$, then, by our assumption, $c \neq b$. But from this it follows that $\rho(\varphi^n b, \varphi^n c) \longrightarrow \infty$ as $n \longrightarrow +\infty$, and since $\rho(\varphi^n c, \varphi^n a) \longrightarrow 0$ as $n \longrightarrow +\infty$, we deduce that $\rho(\varphi^n b, \varphi^n a) \longrightarrow \infty$ as $n \longrightarrow +\infty$. Hence the relations $a \in l_+(b)$ and $a \in l_-(b)$ constitute equivalence relations, and that two curves $l_+(a)$ and $l_+(b)$ (respectively $l_-(a)$ and $l_-(b)$) are either disjoint or coincide.

Finally, it is not difficult to show that the curves $l_+(a)$ and $l_-(a)$ depend continuously on $a \in \mathbf{R}^2$. We shall regard the set of curves l_+ (l_-) as an expanding (contracting) foliation on $\mathbf{R}^2$.

To finish the proof of the Grobman-Hartman theorem, we define mappings $\pi_+ : \mathbf{R}^2 \longrightarrow \gamma_+$ and $\pi_- : \mathbf{R}^2 \longrightarrow \gamma_-$ in the following manner:

$$\pi_+(a) = l_-(a) \cap \gamma_+, \qquad \pi_-(a) = l_+(a) \cap \gamma_-.$$

These mappings are well-defined and continuous (since the families Γ_+ and Γ_- depend continuously upon $a \in \mathbf{R}^2$)* and commute with φ; that is, $\varphi|_{\gamma_+} \circ \pi_+ = \pi_+ \circ \varphi$ and $\varphi|_{\gamma_-} \circ \pi_- = \pi_- \circ \varphi$. Since γ_+ and γ_- are diffeomorphic to $\mathbf{R}^1$ and the mapping $\varphi|_{\gamma_-}$ is a contraction while $\varphi|_{\gamma_+}$ is an expansion, and since the mappings $\varphi|_{\gamma_-}$ and $\varphi|_{\gamma_+}$ are topologically conjugate to linear mappings, it follows that the mapping

$$h: \mathbf{R}^2 \longrightarrow \mathbf{R}^1 \times \mathbf{R}^1, \qquad h(a) = (\pi_+(a), \pi_-(a)),$$

provides the desired homeomorphism between the mapping φ and the linear transformation with matrix $\left(\begin{smallmatrix} 2 & 0 \\ 0 & 1/2 \end{smallmatrix}\right)$. The equivalence of φ with a linear transformation implies all the remaining assertions of the Hadamard-Perron theorem.

REMARKS. 1) Actually, we have obtained somewhat more than what is asserted in the Grobman-Hartman theorem. We have shown the existence of two invariant families of smooth curves depending continuously upon the point $a \in \mathbf{R}^2$.

2) Every hyperbolic linear operator $\widetilde{\varphi}$ in $\mathbf{R}^n$ may be reduced to the form

$$\begin{pmatrix} x \\ y \end{pmatrix} \xrightarrow{\widetilde{\varphi}} \begin{pmatrix} A_+ x \\ A_- y \end{pmatrix},$$

where $x = (x_1, \ldots, x_{k_1})$, $y = (y_1, \ldots, y_{k_2})$, $k_1 + k_2 = n$, A_+ is an expansive linear operator and A_- is a contractive linear operator, and all the proofs given above carry over to the multidimensional case with almost no change in notation.

3) To prove analogous theorems in the case of a flow φ_t, one may either carry out all the preceding arguments directly, or else apply the results obtained above to φ_1, and then prove that the families Γ_+ and Γ_1, invariant with respect to φ_1, are also invariant with respect to φ_t for any t (it is sufficient, however, to prove invariance for any rational t).

**Translator's note.* The author has forgotten to define Γ_+ and Γ_-, but they obviously denote the families of curves l_+ and l_-, respectively.

§6. Stable and unstable manifolds of hyperbolic fixed points. Heteroclinic and homoclinic points

Thus, the Grobman-Hartman and Hadamard-Perron theorems guarantee the existence, in some neighborhood U of a hyperbolic fixed point $x_0 \in M^n$, of two invariant manifolds $W^u_U(x_0)$ and $W^s_U(x_0)$ (here we emphasize the dependence upon the neighborhood U), having the properties that $\varphi W^s_U(x_0) \subset W^s_U(x_0)$ and $\varphi^{-1} W^u_U(x_0) \subset W^u_U(x_0)$. From these local invariant manifolds it is not difficult to obtain the global invariant manifolds

$$W^s(x_0) = \bigcup_{n \leq 0} \varphi^n W^s_U(x_0) \quad \text{and} \quad W^u(x_0) = \bigcup_{n \geq 0} \varphi^n W^u_U(x_0).$$

The manifolds $W^s(x_0)$ and $W^u(x_0)$ are known as the *stable* and *unstable* manifolds (respectively) of the fixed point x_0, and are sometimes called separatrices or "whiskers". Each of these manifolds forms a smooth submanifold of M^n, diffeomorphic to some euclidean space. The restriction of φ to $W^s(x_0)$ is a contraction, and the restriction of φ to $W^u(x_0)$ is an expansion. These manifolds are not closed in M^n, and their closure may have a rather complicated topological structure. These manifolds can be defined in topological terms, and consequently their behavior constitutes a topological invariant of φ.

EXERCISE. Prove that

$$W^s(x_0) = \{x \in M^n | \varphi^n(x) \xrightarrow[n \to +\infty]{} x_0\},$$

$$W^u(x_0) = \{x \in M^n | \varphi^n(x) \xrightarrow[n \to -\infty]{} x_0\}.$$

Examining the proof of the Hadamard-Perron theorem, one can also prove that any compact manifold in $W^s(x_0)$ ($W^u(x_0)$) varies only slightly (in the C^1-topology) for C^1-small variations of φ.

The manifolds $W^s(x_0)$ and $W^u(x_1)$ (or $W^u(x_0)$) may, in general, intersect, and the existence of such a point of intersection implies that the behavior of the trajectories of the diffeomorphism φ is quite complicated.

DEFINITION. The point $x \in M^n$ is said to be a *heteroclinic point* of the diffeomorphism φ if $x \in W^s(x_0) \cap W^u(x_1)$, where x_0 and x_1 are fixed points of φ. The point $x \in M^n$ is said to be a *homoclinic point* of φ if $x \in W^s(x_0) \cap W^u(x_0)$.

DEFINITION. We say that two submanifolds N_1 and N_2 of M^n *intersect transversally* at the point $x \in M^n$ if $TN_1|_x \oplus TN_2|_x = TM^n|_x$.

EXAMPLE 10. We recall the flow φ_t of Example 7b), and assume that the unstable manifold of the upper saddle point intersects with the stable manifold of the lower saddle point; that is, there exists a line of steepest descent leading from one saddle point to the other (see Figure 4). We sketch a portion of the sphere, containing both saddle points, on the plane (Figure 5).

The point x in Figure 5 is a heteroclinic point of the flow φ_t. Now consider the diffeomorphism $\varphi = \varphi_1$. The points $x_0 = F$ and $x_1 = G$ are hyperbolic fixed points of φ, and x is a heteroclinic point; however, the manifolds $W^s(x_0)$ and $W^u(x_1)$ do not intersect transversally at x, and such a heteroclinic point may be eliminated by a small perturbation of φ. We shall perturb φ so that in a neighborhood of x there appears a heteroclinic point which is a transversal intersection of $W^s(x_0)$ and $W^u(x_1)$. For this purpose, we consider a neighborhood V of x which does not intersect any of its own translates; that is, $\varphi^n V \cap V = \varnothing$ (for $n \neq 0$). Let Ω be a C^1-small diffeomorphism of

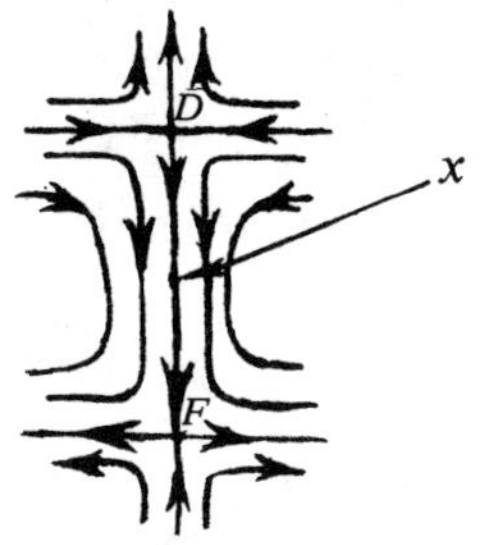

FIGURE 5

M^n which coincides with the identity mapping outside V. We see that one may speak of the stable and unstable manifolds of the diffeomorphism $\Omega \circ \varphi$. It is easily seen that any point $x \in V \cap W^s(x_1)$ belongs to the stable manifold of the fixed point x_1 of $\Omega \circ \varphi$. In fact, the sequence $(\Omega \circ \varphi)^n x$, $n \geqslant 0$, merely coincides with the sequence $\varphi^n x$, $n \geqslant 0$, since $\Omega \circ \varphi$ coincides with φ outside of $\bigcup_{n<0} \varphi^n V$. Clearly, the set $M = \Omega(V \cap W^u(x_0))$ belongs to the unstable manifold of the fixed point x_0 of the diffeomorphism $\Omega \circ f$. In fact,

$$(\Omega \circ \varphi)^{-1} M = (\varphi^{-1} \circ \Omega^{-1}) M = \varphi^{-1}(V \cap W^u(x_0)) \subset W^u(x_0)$$

and

$$(\Omega \circ \varphi)^{-1} M \cap \left(\bigcup_{n \geqslant 0} \varphi^n V \right) = \varnothing \, .$$

Thus, by means of a diffeomorphism which is "concentrated" on V, we can arbitrarily alter the stable manifold of x_0 without changing the unstable manifold of x_1; in particular, we can obtain a point in V at which the stable manifold of x_0 and the unstable manifold of x_1 intersect transversally. In view of the "good" dependence of the stable and unstable manifolds of fixed points upon the diffeomorphism, any diffeomorphism which is close to a diffeomorphism having a transversal heteroclinic point will also have a transversal heteroclinic point.

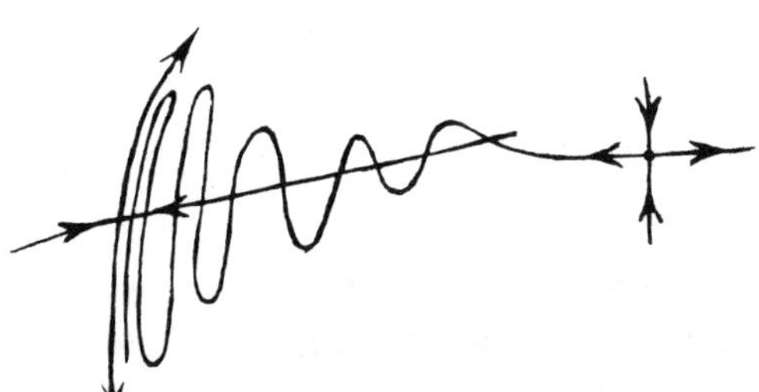

FIGURE 6

Since the trajectory of a transversal heteroclinic point consists of transversal heteroclinic points, the stable and unstable manifolds which pass through a heteroclinic point oscillate strongly and exhibit fairly complicated behavior (see Figure 6), and this behavior constitutes a topological invariant.

EXERCISE. Prove that if $W^s(x_0)$ and $W^u(x_1)$ have a transversal point of intersection, then $W^u(x_0) \subset \overline{W^u(x_1)}$. The situation becomes still more complicated when the diffeomorphism has a transversal homoclinic point.

EXAMPLE 11. We consider a flow φ_t in the plane $\mathbf{R}^2$, having a fixed point x_0 and a separatrix leading from x_0 to x_0 (Figure 7).

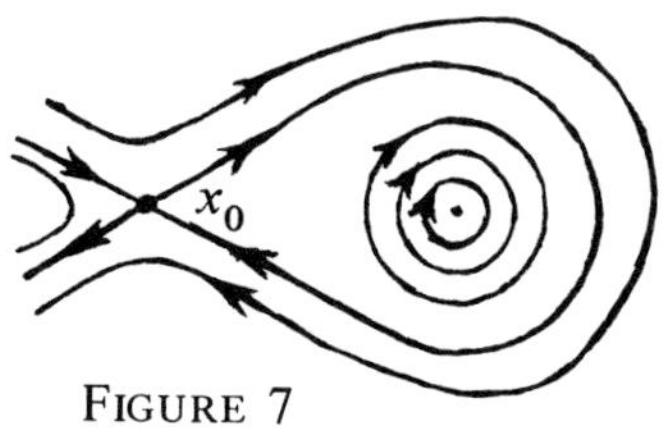

FIGURE 7

Perturbing φ_1 in a manner analogous to that used in Example 10, we obtain a diffeomorphism having a transversal homoclinic point. The existence of such a point implies an extremely complicated behavior of the stable and unstable manifolds as well as the trajectories close to these manifolds (see Figure 8).

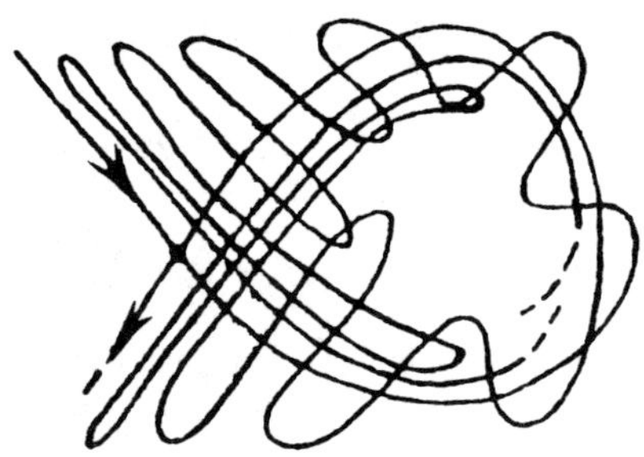

FIGURE 8

EXERCISE. Prove that if $W^u(x_0)$ and $W^s(x_0)$ have a point of transversal intersection, then $W^u(x_0) \subset \overline{W^u(x_0) \backslash K}$, where K is any compact subset of $W^u(x_0)$.

It will be proved below that any neighborhood of a transversal homoclinic point contains periodic points.

It is clear that the notions of stable and unstable manifolds can be carried over to periodic points of a diffeomorphism and even to periodic trajectories of a flow. The notions of homoclinic and heteroclinic points can also be extended in a similar fashion.

§7. Typical properties of dynamical systems

1. DEFINITION. A property of a dynamical system is said to be *typical* provided it is satisfied on the complement of a set of the first category; that is, on the complement of a countable union of closed, nowhere dense sets in the space of dynamical systems.

It follows from the definition that the property of possessing a finite or countable number of typical properties is also a typical property. Any dynamical system can be approximated by a system for which a given typical property is satisfied.

It is clear that whether a property is typical depends upon the space of dynamical systems and the topology assigned to it.

The following (not very complicated) theorem indicates some important generic properties of dynamical systems.

THEOREM OF KUPKA AND SMALE (see [8]). *In the space* $\mathrm{Diff}_k^r(M^n)$, $r \geqslant k \geqslant 1$, *the following three properties are typical*:

G1. *All fixed points are hyperbolic.*

G2. *All periodic points are hyperbolic.*

G3. *Any stable manifold (of a fixed or periodic point) intersects transversally with any unstable manifold.*

Property G1 is even satisfied on an open everywhere dense set. Property G2 includes periodic points with all possible periods. Although, for a fixed period n, the hyperbolicity of periodic points with period n holds on an open everywhere dense set, G2 may be satisfied only on the complement of a set of the second category.*

The proof that G1–G3 are typical, as, indeed, that of almost every assertion concerning "general position," is based upon the Thom transversality theorem ([9], §7) (or later variants of it).

We shall sketch the proof of the fact that G1 is typical, essentially reproducing part of the proof of Thom's theorem.

THEOREM. *For any compact manifold* M^n, *the set of diffeomorphisms in* $\mathrm{Diff}_k^r(M^n)$ *which possess property* G1 *is open and everywhere dense.*

OUTLINE OF THE PROOF. Every hyperbolic fixed point is isolated. In fact, in terms of local coordinates in a neighborhood of the fixed point, the diffeomorphism φ has the form $\varphi(x) = Ax + \cdots$, and the fixed points of φ are determined by the equation $\varphi(x) = x$; that is, $(A - E)x + \cdots = 0$. But since 1 does not lie in the spectrum of A, the operator $A - E$ is invertible, and by the implicit function theorem the equation $(A - E)x + \cdots = 0$ possesses, locally, the unique solution $x = 0$. By the compactness of M^n, it follows that if all the fixed points of φ are hyperbolic, then their number is finite. From the openness of the set of linear hyperbolic operators it follows that, for a small C^1-perturbation, a hyperbolic fixed point is deformed into a hyperbolic fixed point. Hence the set of diffeomorphisms satisfying G1 is open. It remains to prove that this set is everywhere dense.

As is almost always the case for proofs of "general position" theorems, the demonstration is carried out in two stages:

1) First the assertion is proved locally, and this constitutes the heart of the proof.

2) Then, by "pasting together" (i.e., choosing a covering, constructing a partition of unity, etc.) we obtain the final assertion.

---

* *Translator's note.* Presumably, the author meant to say "first category".

We shall confine ourselves to the first stage. Let $\varphi\colon U \longrightarrow \varphi(U) \subset \mathbf{R}^n$ be a local diffeomorphism. We consider the differentiable mapping $\psi\colon U \longrightarrow \mathbf{R}^n$ defined by $\psi(x) = \varphi(x) - x$. The point x_0 is a fixed point of φ if $\psi(x) = 0$; moreover, if $\det(\psi'|_{x_0}) \neq 0$, then the fixed point x_0 is isolated.

DEFINITION. Let $f\colon \mathbf{R}^n \longrightarrow \mathbf{R}^n$ be a differentiable mapping. The point x_0 is said to be a *critical point* of f if $\det(f'|_{x_0}) = 0$. The point $f(x_0)$ is called a *critical value* of f.

We have the following important result.

THEOREM OF SARD. *The set of critical values of any differentiable mapping f has measure zero.*

The proof is based upon the following remark: if on the set $M \subset \mathbf{R}^n$ we have $|\det f'| \leqslant \epsilon$, then $\mu(fM) \leqslant \epsilon\mu(M)$.

Applying Sard's theorem to the mapping ψ, we deduce that, arbitrarily close to 0, there is a point $\vec{c} \in \mathbf{R}^n$ which is not a critical value of ψ. Hence 0 is not a critical value of the mapping $\psi - \vec{c}$, and therefore all fixed points of the mapping $\widetilde{\varphi} = \varphi - \vec{c}$ are isolated. For each fixed point x_i of $\widetilde{\varphi}$, we take a neighborhood V_i of x_i, with $V_i \cap V_j = \varnothing$ for $i \neq j$. By varying the diffeomorphism $\widetilde{\varphi}$ in V_i, we can easily make the fixed point x_i hyperbolic. We leave the completion of the proof to the reader.

Similarly, it can be proved that the hyperbolicity of periodic points with a fixed period holds on an open everywhere dense set, which implies that G2 is typical. Finally, for the proof that G3 is typical, more delicate arguments are necessary. It will be clear from Example 14 that, in general, we cannot expect the set on which G3 is satisfied to be open.

2. We shall now describe yet another property of diffeomorphisms; that it is typical can be proved for any set $\mathrm{Diff}^r(M^n)$, $r \geqslant 1$, but only for the C^1-topology.

DEFINITION. Let X be a topological space and $\varphi\colon X \longrightarrow X$ a homeomorphism. The point $x \in X$ is said to be a *wandering point* of φ if there is a neighborhood U of x which does not intersect any of its own translates: $\forall n \neq 0$, $\varphi^n U \cap U = \varnothing$. Otherwise the point is said to be *nonwandering*. An analogous definition can also be given for a flow.

The set of nonwandering points of a homeomorphism φ is usually denoted by $\Omega(\varphi)$ (or simply Ω), and is obviously a closed set, invariant with respect to φ. Clearly, the behavior of the trajectories of φ is determined, to a great extent, by the structure of the set Ω and the behavior of the trajectories in Ω.

In connection with nonwandering points, there is a certain equivalence relation in spaces of diffeomorphisms and a corresponding notion of stability. Namely, we shall say that the diffeomorphisms f and g are *Ω-conjugate* if there exists a homeomorphism $h\colon \Omega(f) \longrightarrow \Omega(g)$ such that the following diagram is commutative:

$$\begin{array}{ccc} \Omega(f) & \xrightarrow{\;f\;} & \Omega(f) \\ {\scriptstyle h}\big\downarrow & & \big\downarrow{\scriptstyle h} \\ \Omega(g) & \xrightarrow{\;g\;} & \Omega(g) \end{array}$$

The corresponding notion of stability is called *Ω-stability*. Clearly, topological conjugacy implies Ω-conjugacy, and therefore roughness implies Ω-stability. The converse assertion is not valid on an open set of diffeomorphisms (see Example 14 below and the theorem on Ω-stability in §6 of the lectures of A. B. Katok contained in this volume).

EXERCISE. Prove that a nonwandering point cannot be stable in the sense of Poisson.

We can now formulate the following property:

G4. The set of fixed and periodic points is everywhere dense in the set of nonwandering points.

Pugh [10] proved that property G4 is typical for $\mathrm{Diff}^r(M^n)$ in the C^1-topology; in particular, any diffeomorphism may be C^1-approximated by a diffeomorphism for which G4 holds. For any compact M^n and any diffeomorphism φ, the set $\Omega(\varphi)$ of nonwandering points is nonempty, and consequently a diffeomorphism which satisfies property G4 has periodic points.

EXERCISE. Prove that no C^1-stable diffeomorphisms exist on any manifold M^n.

At the present time, it is not known whether property G4 is typical in the C^2-topology. In particular, we do not know whether, for any manifold M^n, the set of diffeomorphisms having periodic points is everywhere dense in $\mathrm{Diff}^2_2(M^n)$.

The theorem that G4 is typical in the C^1-topology may be deduced in a comparatively simple manner from the following difficult theorem, proved by Pugh [11] and known as the

CLOSING LEMMA. *Let x be a nonwandering point of the diffeomorphism φ. Then there exists an arbitrarily small (in the C^1-topology) perturbation of φ having x as a periodic point.*

Clearly, properties G1–G4 and the closing lemma may be reformulated for flows:

G1. All fixed points are hyperbolic.

G2. All periodic trajectories are hyperbolic.

G3. Any stable manifold (of a fixed point or a periodic trajectory) intersects any unstable manifold transversally.

G4. The set of fixed points and periodic trajectories is everywhere dense in the set of nonwandering points.

The closing lemma, and the fact that G4 in the C^1-topology and G1–G3 in $\mathcal{X}^r$ are typical, remain valid for flows.

If G1–G4 are satisfied for a flow on a compact two-dimensional manifold, then the following assertions hold:

1) The number of fixed points and periodic trajectories is finite (and their eigenvalues satisfy a certain inequality which guarantees hyperbolicity).

2) There exist no trajectories (separatrices) leading from a saddle point to a saddle point.

3) The ω- and α-limit sets of any trajectory are contained in the union of the fixed points and the periodic trajectories (limit cycles).

In 1937, Andronov and Pontrjagin [1] proved (for flows on a disk) that conditions 1)–3) are necessary and sufficient for roughness (topological stability), and later Peixoto [2], [3] proved the same assertion for an arbitrary two-dimensional manifold. The flow of Example 7a) satisfies conditions 1)–3), and consequently is topologically stable.

§8. New examples of dynamical systems

1. Thus, for flows on two-dimensional surfaces, we have a very reassuring picture. We have simple geometrical conditions which are necessary and sufficient for roughness, and rough systems are dense among all systems. Moreover, it is easy to enumerate the possible types of behavior of the trajectories of a rough system.

As we have already mentioned, everything becomes far more complicated in higher dimensions. But before turning to new examples, we shall describe certain multi-dimensional analogues of rough flows on two-dimensional manifolds.

DEFINITION. By a *Morse-Smale diffeomorphism,* we mean a diffeomorphism $\varphi\colon M^n \longrightarrow M^n$ satisfying the following conditions:

1) The set $\Omega(\varphi)$ of nonwandering points is finite (and consequently consists only of fixed and periodic points).

2) All fixed and periodic points are hyperbolic.

3) Any stable manifold (of a fixed or periodic point) intersects any unstable manifold transversally.

Around 1960 Smale conjectured that the set of diffeomorphisms satisfying conditions 1)–3) is everywhere dense,([1]) and also conjectured that these diffeomorphisms are structurally stable. The former conjecture was subsequently shown to be false, while the conjecture regarding roughness (topological stability) was proved quite recently (1968).

THEOREM OF PALIS AND SMALE [12]. *The diffeomorphisms satisfying conditions* 1)–3) *form an open set, and are structurally stable.*

In proving openness, the most difficult point is verifying that property 1 is preserved under perturbations. This verification requires consideration of the limit behavior of all the trajectories. Property 3) is absolutely necessary, since without it the finite set $\Omega(\varphi)$ could easily become infinite through a perturbation of φ. Such a situation is known as an *Ω-explosion.* Broadly speaking, the interest of this theorem stems from the fact that it copes with heteroclinic points and the concomitant complicated behavior of stable and unstable manifolds. (This theorem can be extended, in an obvious fashion, to the case of flows, and, in an entirely nonobvious fashion, to the far more complicated situation of diffeomorphisms φ for which the set $\Omega(\varphi)$ is infinite. See §6 of the lectures of A. B. Katok contained in this volume.)

([1]) The Lefschetz fixed point theorem easily implies that for some manifolds diffeomorphisms belonging to a certain homotopy class automatically have an infinite number of periodic points. Therefore density has to be interpreted here, for example, as density among diffeomorphisms close to the identity diffeomorphism.

Using the Palis-Smale theorem, we can now construct rough dynamical systems on any compact manifold M^n.

PROBLEM. Let M^n be a smooth compact Riemannian manifold. Then, for an open, everywhere dense set of smooth functions $f: M^n \longrightarrow \mathbf{R}^1$, the flow φ_t, defined by the equation $\dot{x} = \operatorname{grad} f$, satisfies conditions 1)–3) and is therefore structurally stable.

2. There exist structurally stable systems with far more complicated behavior than Morse-Smale systems. The theory of such systems has just arisen within the last ten years. We now turn to the construction of examples which played a major role in the inception of this theory.

EXAMPLE 12. We consider the two-dimensional torus T^2, which may be defined as the factor group of $\mathbf{R}^2$ by the integral lattice $\mathbf{Z} \oplus \mathbf{Z}$, i.e. $T^2 = \mathbf{R}^2/\mathbf{Z} \oplus \mathbf{Z}$. We denote the natural projection of $\mathbf{R}^2$ onto this group by $\pi: \mathbf{R}^2 \longrightarrow T^2$. Consider the automorphism $\widetilde{\varphi}: \mathbf{R}^2 \longrightarrow \mathbf{R}^2$ defined by the matrix $A = \left(\begin{smallmatrix} 2 & 1 \\ 1 & 1 \end{smallmatrix}\right)$. Since the matrix A is integral and unimodular ($|\det A| = 1$), $\widetilde{\varphi}$ induces an automorphism of $\mathbf{Z} \oplus \mathbf{Z}$ and therefore induces an automorphism φ of $\mathbf{R}^2/\mathbf{Z} \oplus \mathbf{Z}$. Thus we have constructed a diffeomorphism $\varphi: T^2 \longrightarrow T^2$. The linear transformation $\widetilde{\varphi}$ is a hyperbolic bending map and has two eigendirections, along one of which a contraction occurs, and along the other an expansion. Figure 9 shows the image of the coordinate square under the

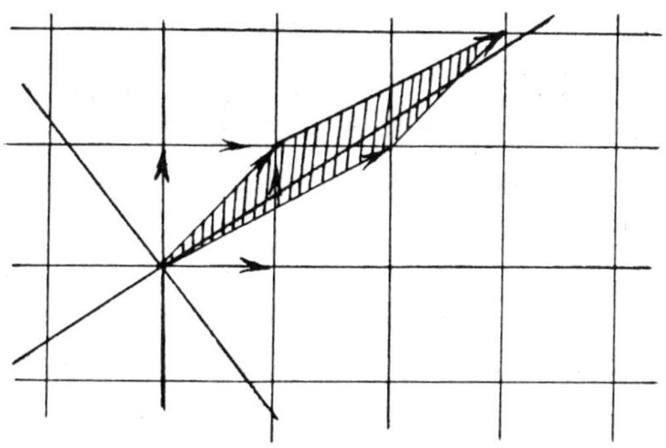

FIGURE 9

action of $\widetilde{\varphi}$ and the invariant lines of $\widetilde{\varphi}$. Under the action of the projection π, the origin goes over into a hyperbolic fixed point x_0 of φ, and the invariant lines go over into the stable and unstable manifolds of x_0. Each of these manifolds is dense in T^2. The families of lines invariant with respect to $\widetilde{\varphi}$, parallel with each of the eigendirections, are projected by π into families of manifolds in T^2 which are invariant with respect to φ. We shall refer to these families as *foliations* of T^2. The images of the lines parallel with the stable manifold of the mapping $\widetilde{\varphi}$ will be called *contracting leaves,* and the images of the lines parallel with the unstable manifold of $\widetilde{\varphi}$ will be called *expanding leaves.* Each expanding or contracting leaf is dense in T^2. For any two points belonging to one of the contracting leaves, the distance between their images under φ^n approaches zero as $n \longrightarrow +\infty$. Similarly, points lying in one of the expanding leaves converge under the action of φ^n as $n \longrightarrow -\infty$. The rational points of the torus T^2 (that is, the images under the mapping π of the rational points of the plane) are periodic points of φ. In fact, the number of rational points with a fixed denominator q is equal to q^2, and the set of such points is mapped by φ into itself. Conversely, it is easily shown that every periodic point of φ is rational. Thus the periodic points of φ are dense in T^2. The stable (or unstable) manifold of any periodic point is dense in T^2, and therefore every trajectory is unstable: in an arbitrarily small neighborhood of any given point, one can find points whose trajectories have different asymptotic behavior.

EXERCISE. Show that there exists a point $x \in T^2$ such that the trajectory passing through x is everywhere dense. Show that the trajectory of almost every point is everywhere dense.

It turns out that the diffeomorphism φ is topologically stable (rough); that is, for any diffeomorphism ψ which is C^1-close to φ, there exists a homeomorphism h, close to the identity, such that $\varphi \circ h = h \circ \psi$.

EXERCISE. Let h be a continuous mapping of T^2 into itself which is close to the identity mapping of T^2 and satisfies the condition $\varphi \circ h = h \circ \varphi$. Prove that h is the identity mapping of T^2.

3. The diffeomorphism φ constitutes a representative of an important class of dynamical systems, the so-called Y-systems (or Anosov flows), which were introduced and studied in detail by D. V. Anosov in [13] (see Definition 1.3 in the lectures of A. B. Katok contained in this volume). In particular, he proved that any Y-diffeomorphism (Anosov diffeomorphism) is structurally stable. Several proofs of this theorem (or, at any rate, several different presentations of the proof) are known at the present time. We shall sketch a proof which is close to the original proof of Anosov, and also carry out more rigorously certain arguments close to the arguments of Moser. Finally, still another proof will be presented in Katok's lectures.

Anosov's proof of the roughness of a Y-diffeomorphism is based upon a line of reasoning which goes back as far as Perron, and is known as the theorem on constantly acting perturbations. There exist many variants of this theorem. We shall describe one of the simplest, which we shall need in proving the roughness of φ.

THEOREM ON CONSTANTLY ACTING PERTURBATIONS. *Let there be given a sequence* $\{\mathbf{R}_i^n\}_{-\infty<i<\infty}$ *of copies of one and the same space* $\mathbf{R}^n$, *and a sequence of identical hyperbolic linear transformations* $\varphi_i\colon \mathbf{R}_i^n \longrightarrow \mathbf{R}_{i+1}^n$. *In each of the spaces* $\mathbf{R}_i^n$, *consider a diffeomorphism* g_i, *ϵ-close to the identity in the* C^1-*norm* (*where* ϵ *is independent of* i). *Finally, in each of the* $\mathbf{R}_i^n$, *fix an identical neighborhood* V_i *of the origin. Let* $R = \bigcup_i \mathbf{R}_i^n$, $V = \bigcup_i V_i$ *and* $f\colon R \longrightarrow R$.* *Then, for sufficiently small* ϵ, *there exist, in every neighborhood* V_i,

a) *a* C^1-*submanifold* W_i^s, *consisting of those and only those points of* V_i *which always remain in* V *under the application of the transformations* $(f \circ g)^n$, $n = 0, 1, 2, \ldots$, *and*

b) *a* C^1-*submanifold* W_i^u, *consisting of those and only those points of* V_i *which always remain in* V *under the application of the transformations* $(g^{-1} \circ f^{-1})^n$, $n = 0, 1, 2, \ldots$.

**Translator's note.* The author does not define f and g, but presumably they are the maps defined by the conditions $f|_{\mathbf{R}_i} = f_i$ and $g|_{\mathbf{R}_i} = g_i$.

Moreover,

c) *the manifolds* W_i^s *and* W_i^u *intersect transversally in a unique point* x_i, *and*

d) *the* C^1*-submanifolds* W_i^s *and* W_i^u, *as well as the point* x_i, *depend continuously upon* g.

OUTLINE OF THE PROOF. Let $n = 2$, let f_i be defined by $(x, y) \longrightarrow (2x, y/2)$, and V_i by $|x|, |y| < 1$; we consider the space $\mathbf{R}_0^2$ and seek the set of those points of $\mathbf{R}_0^2$ which remain in V under the action of f^n, $n = 0, 1, 2, \ldots$. Those points of $\mathbf{R}_0^2$ which remain in V after the application of $f, f^2, \ldots, f^k$ obviously belong to $V_0 \cap f^{-1}(V_1) \cap \cdots \cap f^{-k}(V_k)$. Thus we obtain in V_0 a nested sequence of rectangles $|y| < 1$, $|x| < (\frac{1}{2})^k$, the intersection of which is just the segment $|y| < 1$, $x = 0$. We have considered merely the case where g is the identity mapping. In case g is not the identity, the sets $V_0 \cap (g \circ f)^{-1}(V_1) \cap \cdots \cap (g \circ f)^{-k}(V_k)$ will no longer be rectangles, but rather sets bounded by two curves, both of which are C^1-close (uniformly in k) to a vertical line; moreover, with increasing k, the distance between these curves will exponentially converge to zero, and the intersection of the inverse images of V_k under $(g \circ f)^k$ will be a C^1-curve, close to a vertical line. Similarly, one constructs the set of points which remain in V under the action of $(g \circ f)^n$, $n = 0, -1, -2, \ldots$, and this set consists of a nearly horizontal C^1-curve. The two curves so constructed clearly intersect in a unique point x_0. From the construction of the submanifolds W_0^s and W_0^u it is not difficult to deduce that these submanifolds depend continuously upon g_i for any fixed i, and depend very weakly upon g_i for sufficiently large $|i|$. More precisely, for any $\delta > 0$ there is an $N = N(\delta)$ such that any admissible variation of all the g_i having indices greater (in absolute value) than N will give rise to a variation of W_0^s and W_0^u which does not exceed δ in the C^1-norm. When carrying out a precise proof of the existence of W_0^u and W_0^s, one may utilize the arguments which we used in proving the Grobman-Hartman theorem; that is, the construction of contracting and expanding foliations of $\mathbf{R}^2$.

Proof of the roughness of the diffeomorphism φ. Let g be a diffeomorphism of T^2 which is C^1-close to the identity. We must prove that φ and $g \circ \varphi$ are topologically equivalent. Suppose there exists a homeomorphism h which is ϵ-close to the identity and such that $\varphi = h^{-1} \circ (g \circ \varphi) \circ h$. Then, for any $n \in \mathbf{Z}$, $\varphi^n = h^{-1} \circ (g \circ \varphi)^n \circ h$, and so for any point x the trajectory of hx under the action of $g \circ \varphi$ lies in the ϵ-neighborhood of the trajectory of x under the action of φ. In fact, $h(\varphi^n x) = (g \circ \varphi)^n(hx)$. Conversely, using the theorem on constantly acting perturbations, we can show that for any point $x \in T^2$ there exists a unique point x' whose trajectory under the action of $\varphi \circ g$ lies in the ϵ-neighborhood of the trajectory of x under the action of φ, and, moreover, that x' depends continuously upon x. Setting $hx = x'$, we thus obtain a continuous mapping $h\colon T^2 \longrightarrow T^2$ which is close to the identity mapping (of the compact manifold T^2 into itself) and is therefore an "onto" mapping. It follows from the uniqueness of x' that $h \circ \varphi = \psi \circ h$, where $\psi = g \circ \varphi$. Multiplying this equation on the left by ψ, we obtain $(\psi \circ h) \circ \varphi = \psi^2 \circ h$, or $(h \circ \varphi^2) = \psi^2 \circ h$. Similarly, we obtain $h \circ \varphi^n = \psi^n \circ h$ for any $n \geqslant 1$. Finally, multiplying the latter equation on the left by ψ^{-n} and on the right by φ^{-n}, we obtain $\psi^{-n} \circ h = h \circ \varphi^{-n}$. Thus $h \circ \varphi^n = \psi^n \circ h$ for any $n \in \mathbf{Z}$.

As we shall now prove, this relation implies that h does not transform any two points into one, and is therefore a homeomorphism. In fact, since h is ϵ-close to the identity mapping of T^2, any two points x_1 and x_2 which satisfy the condition $\rho(x_1, x_2) > 2\epsilon$ have distinct images. Now, assume that the points x_1 and x_2 satisfy the condition $\rho(\varphi^n x_1, \varphi^n x_2) > 2\epsilon$ for some n. Then the points $h\varphi^n x_1$ and $h\varphi^n x_2$ are distinct; but then it follows from the equality $h \circ \varphi^n = \psi^n \circ h$ that hx_1 and hx_2 are also distinct. However, it is not difficult to see that for sufficiently small $\epsilon > 0$ the following property holds: for any two distinct points $x_1, x_2 \in T^2$ there exists an n such that $\rho(\varphi^n x_1, \varphi^n x_2) > 2\epsilon$. Thus, h does not identify any two distinct points and is therefore a homeomorphism.

Thus, it remains to show how the theorem on constantly acting perturbations can be used to construct the continuous mapping h. Let ψ be a diffeomorphism of T^2 which is C^1-close to φ; then $\psi = g \circ \varphi$, where g is a diffeomorphism of T^2 which is C^1-close to the identity. The diffeomorphism g induces a diffeomorphism $\widetilde{g}: \mathbf{R}^2 \longrightarrow \mathbf{R}^2$ which is periodic with period 1 in x and y. We consider the point $x \in T^2$,* and choose some representative $\widetilde{x}$ of x in $\mathbf{R}^2$. Now let $\mathbf{R}^2_n$ be the set of vectors in $\mathbf{R}^2$ with initial point at $\widetilde{\varphi}^n\widetilde{x}$. The mappings $\widetilde{\varphi}$ and $\widetilde{g}$ induce mappings φ_i and g_i which satisfy all the conditions of the theorem on constantly acting perturbations.

4. Second proof of the roughness of the diffeomorphism φ and the Grobman-Hartman theorem. Consider the following three mappings of $\mathbf{R}^2$ into itself:

$$F: z \longmapsto Az,$$

$$G: z \longmapsto Az + g(z),$$

$$H: z \longmapsto z + h(z),$$

where $z = (x, y)$, $A = \begin{pmatrix} 2 & 1 \\ 1 & 1 \end{pmatrix}$, the function

$$g(z) = \begin{pmatrix} g_1(x, y) \\ g_2(x, y) \end{pmatrix}$$

is a differentiable vector-valued function, and the function

$$h(z) = \begin{pmatrix} h_1(x, y) \\ h_2(x, y) \end{pmatrix}$$

is a continuous vector-valued function. Writing the equation

$$H \circ F = G \circ H \tag{$*$}$$

in terms of A, g and h, we obtain

$$h(Az) - Ah(z) = g(z + h(z)).$$

**Translator's note.* The author has unfortunately used the same letter x to represent both abscissas in $\mathbf{R}^2$ and points in T^2.

Observe that the left-hand side is linear in h, and denote it by Lh. Then the equation takes the form $Lh(z) = g(z + h(z))$. Thus, if we wish to prove that mappings F and G of the space $\mathbf{R}^2$ into itself are topologically equivalent, then we must, for a fixed vector-valued function g, solve the preceding equation for h. We shall solve this equation by means of the following sequence of approximations:

$$h_0 = 0, \quad h_{n+1}(z) = L^{-1}g(z + h_n(z)).$$

We shall show that the operator L is invertible in the space of continuous functions and that L^{-1} is bounded. Consequently, the sequence h_n will converge uniformly under the condition $\|L^{-1}\| \cdot \sup_z g' < 1$. To prove the invertibility of L and estimate the norm of L^{-1}, consider the eigenvectors of the matrix A:

$$Ae_1 = \lambda_1 e_1, \quad Ae_2 = \lambda_2 e_2, \quad 0 < \lambda_1 < 1 < \lambda_2.$$

Any vector-valued function $h(z)$ is uniquely representable in the form $f_1(z)e_1 + f_2(z)e_2$, where $f_1(z)$ and $f_2(z)$ are scalar-valued functions. The operator L preserves this decomposition:

$$Lh(z) = [f_1(Az) - \lambda_1 f_1(z)]\, e_1 + [f_2(Az) - \lambda_2 f_2(z)]\, e_2;$$

that is, L transforms the subspace of vector-valued functions proportional to e_i into itself. Thus it suffices to show that the operators L_i: $f(z) \longrightarrow f(Az) - \lambda_i f(z)$ are invertible in $C(\mathbf{R}^2)$. The spectrum of the operator $\mathsf{A} : f(z) \longrightarrow f(Az)$ is confined to the unit disk, and $|\lambda_i| \neq 1$; therefore the operators L_i are invertible. In fact, A is invertible and $\|\mathsf{A}\| = \|\mathsf{A}^{-1}\| = 1$; hence

$$L_1^{-1} = (\mathsf{A} - \lambda_1 E)^{-1} = [(E - \lambda_1 \mathsf{A}^{-1})\mathsf{A}]^{-1} = \mathsf{A}^{-1}[E - \lambda_1 \mathsf{A}^{-1}]^{-1}$$
$$= \mathsf{A}^{-1}[E + \lambda_1 \mathsf{A}^{-1} + (\lambda_1 \mathsf{A}^{-1})^2 + \cdots]$$

and

$$\|L_1^{-1}\| \leqslant \|\mathsf{A}^{-1}\| \cdot (1 + \|\lambda_1 \mathsf{A}^{-1}\| + \cdots) = 1 \cdot \frac{1}{1 - \lambda_1}.$$

The invertibility of L_2 may be proved in a similar fashion.

To prove the convergence of the successive approximations, we estimate

$$|h_{n+1} - h_n| = |L^{-1}[g(z + h_n(z)) - g(z + h_{n-1}(z))]|$$
$$\leqslant \|L^{-1}\| \cdot \sup_z \left\|\frac{\partial g}{\partial z}\right\| \cdot |h_n(z) - h_{n-1}(z)|.$$

Thus, if the derivatives of the vector-valued function g are sufficiently small, then the successive approximations converge, and the equation $(*)$ has a solution in the class of continuous mappings of $\mathbf{R}^2$. Finally, arguing as in the first proof of the roughness of φ, we deduce that $(*)$ implies that H is a homeomorphism. Up to now, we have been speaking only of mappings of $\mathbf{R}^2$. However, if $g(z)$ and $h(z)$ are vector-valued functions which are periodic in x and y, then F, G and H may be interpreted as mappings of T^2, and it is easily verified that none of the preceding constructions leave the class of periodic functions.

REMARK 1. It is clear that in the preceding proof we have not actually made use of the fact that the dimension is 2 or of the special form of the matrix A. All the arguments carry over without change to any hyperbolic matrix A in the case of mappings of $\mathbf{R}^n$, and for any hyperbolic integral unimodular matrix in the case of mappings of $T^n = \mathbf{R}^n/\mathbf{Z}^n$ (see Example 1.3 in the lectures of A. B. Katok contained in this volume).

REMARK 2. Similar arguments are also applicable to a hyperbolic fixed point of a diffeomorphism of a Banach space (an invertible operator in a Banach space is said to be *hyperbolic* if its spectrum does not intersect the unit circle).

5. EXAMPLE 13. Let $M^3 = T^2 \times \mathbf{R}$. We define a diffeomorphism f of the open manifold M^3 into itself by the following formula:

$$f(z, t) = (\varphi z, t/10), \qquad z \in T^2, \quad t \in \mathbf{R},$$

where φ is the diffeomorphism of Example 12. The diffeomorphism f has the invariant two-dimensional manifold $T^2 \times \{0\}$, and $f|_{T^2 \times \{0\}} = \varphi$. Repeating the arguments given above, one may show that f is topologically stable; in particular, any diffeomorphism which is C^1-close to f has a unique compact invariant manifold which is C^0-close to $T^2 \times \{0\}$. One can also prove that the invariant manifold so obtained is, in fact, C^1-smooth and C^1-close to $T^2 \times \{0\}$. "Multiplying" each leaf of the contracting foliation of the diffeomorphism $f|_{T^2 \times \{0\}}$ by $\mathbf{R}$, we obtain a two-dimensional foliation of $T^2 \times \mathbf{R}$, and we shall denote by $W^s(x)$ the leaf of this foliation which passes through the point x. A leaf of the foliation W^s may be defined in purely topological terms; in fact, it is easily seen that $x_2 \in W^s(x_1)$ if and only if $\rho(f^n x_1, f^n x_2) \longrightarrow 0$ as $n \longrightarrow +\infty$, and that this condition does not depend upon the choice of the metric ρ. As we have actually shown when considering hyperbolic points in the proof of the Grobman-Hartman theorem, for any diffeomorphism which is C^1-close to f there exists an invariant foliation whose leaves have C^1 smoothness and depend continuously upon the perturbation of f. We shall say that a leaf of the foliation W^s is *rational* if it contains at least one periodic point of $f|_{T^2 \times \{0\}}$, and *irrational* otherwise. The set of rational leaves (and, likewise, the set of irrational leaves) is dense in M^3. Since f is topologically stable and, moreover, the appropriate conjugating homeomorphism is unique in the class of homeomorphisms which are close to the identity, the notion of rational and irrational leaves carries over to any C^1-small perturbation of f, and the assertion concerning density of the rational and irrational leaves is valid for any such perturbation.

On the basis of Example 13, we shall now construct an open set of nonrough diffeomorphisms.

THEOREM (SMALE [4]). *The set of C^1-diffeomorphisms of the three-dimensional torus T^3 contains an open set U of diffeomorphisms such that for any $f \in U$ the set of diffeomorphisms which are not homeomorphic to f is dense in U.*

We shall construct the set U as a small neighborhood of a certain diffeomorphism $f \in \mathrm{Diff}(T^3)$, which we now construct.

EXAMPLE 14 (SMALE). Let $\varphi: T^2 \longrightarrow T^2$ be the diffeomorphism of Example 12 and let $l: S^1 \longrightarrow S^1$ be a diffeomorphism of the circle having two fixed points 0 and π, such that the eigenvalues of l at 0 and π are equal to 1/10 and 10, respectively. Consider the diffeomorphism $f_1: T^2 \times S^1 \longrightarrow T^2 \times S^1$, defined as the direct product of φ and l. The diffeomorphism f_1 has the two invariant submanifolds $T^2 \times \{0\}$ and $T^2 \times \{\pi\}$. Let x_0 denote the fixed point of φ, and modify f_1 in a small neighborhood of the fixed point $x_0 \times \pi$ in such a manner that $x_0 \times \pi$ becomes a hyperbolic fixed point of the new diffeomorphism f_2, with unstable manifold $x_0 \times \{S^1 \backslash 0\}$. Since f_2 coincides with f_1 in a neighborhood of $T^2 \times \{0\}$, the restriction of f_2 to some neighborhood of $T^2 \times \{0\}$ is diffeomorphic to the restriction of the diffeomorphism f of Example 13 to some neighborhood of $T^2 \times \{0\} \subset T^2 \times \mathbf{R}$. In particular, a stable foliation W^s of the diffeomorphism f_2 is defined in a neighborhood of $T^2 \times \{0\}$; moreover, the leaves of this foliation can be defined in topological terms. Consider a point x_1, sufficiently close to $T^2 \times \{0\}$, which belongs to the unstable manifold of the fixed point $x_0 \times \pi$, and a small neighborhood V of x_1, which does not intersect its own transforms under the action of f_2. The neighborhood V is fibered into the connected components of its intersection with leaves of the foliation W^s, and, as follows from the construction of f_2, the intersection of V with the unstable manifold of the fixed point $x_0 \times \pi$ lies entirely in such a connected component. Consider a diffeomorphism Ω which is "concentrated" in V, and set $f_3 = \Omega \circ f_2$ (we have carried out a similar construction in Example 10). From the definition of f_3 it is clear that the foliations W^s for f_2 and f_3 will coincide in V, provided the intersection of V with the unstable manifold of $x_0 \times \pi$ for f_3 arises from the corresponding intersection for f_2 by the application of Ω. We choose Ω in such a manner that the unstable manifold of $x_0 \times \pi$ with respect to f_3 has at x_1 one-sided tangency with the connected component of the intersection with V of the leaves of the foliation W^s, and we also require that other tangencies do not occur in the neighborhood V. We assert that some neighborhood U of f_3 in $\mathrm{Diff}(T^3)$ satisfies the requirements in the conclusion of Smale's theorem.

PROOF. We shall say that a smooth two-dimensional manifold in T^3 is *topologically tangent* to a smooth one-dimensional manifold if there is a point x_1 belonging to both manifolds and such that in a certain neighborhood of x_1 the one-dimensional manifold lies on one side of the two-dimensional manifold. Clearly, this notion of tangency is purely topological and is preserved by homeomorphisms (obviously every point of topological tangency is also an ordinary point of tangency, but the converse is not true).

Since, in Example 13, the leaves of the stable foliation W^s depend continuously upon f, the leaves of W^s for diffeomorphisms close to f_3 will depend continuously upon the perturbation. As we have already pointed out, the hyperbolic fixed point $x_0 \times \pi$ and its unstable manifold also depend continuously upon the perturbation. Hence for all diffeomorphisms which are C^1-close to f_3 (we denote the set of such diffeomorphisms by U) there is in some neighborhood of x_1 a point of tangency between the unstable manifold of the fixed point close to $x_0 \times \pi$ and some leaf of W^s. Clearly, the rationality or irrationality of that leaf of W^s which is tangent to the one-dimensional unstable manifold of the fixed point is invariant under topological conjugacy. But for any diffeomorphism $f \in U$, both the rational and irrational leaves of W^s are everywhere dense, and by perturbing f in V we can, without changing W^s, arbitrarily alter the unstable manifold of the fixed point. Therefore, if for the diffeomorphism f the tangency occurs with a rational leaf of W^s, then, by an arbitrarily small perturbation, one may arrange matters so that the tangency occurs with an irrational leaf of W^s, and conversely.

Using the Kupka-Smale theorem, we now show that for any $f \in U$ the set of diffeomorphisms which are not homeomorphic to f is dense in U. In fact, the presence of a topological tangency between the stable manifold of some periodic point and the unstable manifold of some fixed point constitutes a topological invariant. By the Kupka-Smale theorem, for an everywhere dense set of diffeomorphisms there is no point of tangency, and therefore no point of topological tangency between the unstable manifold of some fixed point and the stable manifold of some periodic point (and every rational leaf is such a manifold). On the other hand, as was shown above, diffeomorphisms having a point of tangency between the stable manifold of some periodic point and the unstable manifold of the fixed point close to $x_0 \times \pi$ are everywhere dense in U. This completes the proof.

6. The two examples which follow below are Y-flows in the sense of Anosov.

EXAMPLE 15. Consider the diffeomorphism of T^2 given in Example 12 and the suspension of this diffeomorphism, which is a flow φ_t on the three-dimensional manifold M^3. The flow φ_t is rough (structurally stable), i.e. every flow which is C^1-close to φ_t will be topologically equivalent to φ_t. The flow φ_t has no fixed points, but has a countable number of hyperbolic periodic trajectories with increasing periods. The periodic trajectories of φ_t are dense in M^3.

EXAMPLE 16 (a geodesic flow on a surface of constant negative curvature). It is well known that one can introduce a metric with constant negative curvature on a surface M^2 of genus 2. Consider the manifold M^3 formed by the unit tangent vectors to M^2. We define a certain flow φ_t on M^3, as follows. Let e be a unit tangent vector with origin at the point $x \in M^2$. There is a unique geodesic through x in the direction of e. Displacing this vector along this geodesic through a distance t, we obtain a vector which we denote by $\varphi_t(e)$.

The flow φ_t is, in many respects, similar to the diffeomorphism φ of Example 12. For any point $x \in M^3$, the set of those points which approach the trajectory of x as $t \longrightarrow +\infty$ forms a two-dimensional submanifold $W^s(x)$ which is dense in M^3, and likewise the set of those points which approach the trajectory of x as $t \longrightarrow -\infty$ forms a two-dimensional submanifold $W^u(x)$ which is also dense in M^3. Using the theorem on constantly acting perturbations, one can prove that the flow φ_t is structurally stable.

This flow exhibits fairly complicated behavior; for example, the periodic trajectories are everywhere dense in M^3 (this means that every geodesic segment on M^2 can be approximated by a segment of a closed geodesic).

Appendix. Dynamical systems with time G

Let G be an abstract group and M^n a smooth manifold. A homomorphism φ: $G \longrightarrow \mathrm{Diff}(M^n)$ of G into the group of diffeomorphisms of M^n is called an *action* of G on M^n. Alternatively, one may say that an action of G on M^n is a mapping φ: $G \times M^n \longrightarrow M^n$ which is differentiable with respect to the second argument and satisfies the following relations:

1) $\varphi(e, x) \equiv x$,
2) $\varphi(g_1, \varphi(g_2, x)) = \varphi(g_1 g_2, x)$.

In the important special case where G is a Lie group, we shall, in addition, require differentiability of φ with respect to the first argument.

In the case $G = \mathbf{Z}$ ($G = \mathbf{R}$), we obtain the ordinary definition of a dynamical system with discrete (continuous) time.

In the case of a system $\dot{z} = f(z)$ of ordinary differential equations with analytic right-hand side, we may regard the time z as a complex number and so obtain, under certain additional assumptions, a dynamical system with time $\mathbf{C}$.

EXAMPLES. 1. $G = \mathbf{Z}_2$. The action of such a group is determined by prescribing an involutory diffeomorphism of M^n. For example, if $M^n = S^n = \{x: x_0^2 + \cdots + x_n^2 = 1\}$, then the following diffeomorphisms are involutory:

$$(x_0, x_1, \ldots, x_n) \mapsto (-x_0, -x_1, \ldots, -x_k, x_{k+1}, \ldots, x_n).$$

2. $G = S^1 = \{z: |z| = 1\}$ and $M^n = S^3 = \{(z_1, z_2): |z_1|^2 + |z_2|^2 = 1\}$. The action is defined in the following manner:

$$\varphi(z, (z_1, z_2)) = (z^m z_1, z^n z_2), \qquad m, n \in \mathbf{Z}.$$

3. $G = \mathrm{SO}(3)$, the group of orthogonal matrices of order 3 with determinant 1. This group acts naturally on $\mathbf{R}^3$, on S^2, and also on the set of uniform polynomials of degree k on $\mathbf{R}^3$.

4. $G = \mathbf{Z} \times \mathbf{Z}$. The action of such a group on M^n is defined by a pair of diffeomorphisms $f: M^n \longrightarrow M^n$, $g: M^n \longrightarrow M^n$ which commute with each other ($f \circ g = g \circ f$), and $(m, n)x = (f^m \circ g^n)(x)$.

For dynamical systems with time G, the same problems present themselves as in the cases $\mathbf{R}$ and $\mathbf{Z}$. But, in addition, new difficulties arise in connection with the fact that the space of actions of a fixed group G upon a fixed manifold M^n may have, even locally, a very complicated structure.

Let φ: $G \times M^n \longrightarrow M^n$ be an action of a group G on a manifold M^n. We shall often write gx in place of $\varphi(g, x)$.

DEFINITIONS. By the *orbit* of the point x, we mean the set $Gx = \{y: y = gx, g \in G\}$. The *stationary subgroup* of the point x is the subgroup $H_x = \{g \in G: gx = x\}$. A *fixed point* is a point x_0 such that $Gx_0 = G$; that is, $\forall g \in G$, $gx_0 = x_0$.

Clearly, the orbit of the point x is isomorphic to the coset space G/H_x.

In the case where G is a connected simply connected Lie group, a differentiable action of G on M^n can be defined by means of a homeomorphism ψ of the Lie algebra of G into the Lie algebra $\Gamma(TM^n)$ of vector fields on M^n. To say that the mapping ψ is *homomorphic* means, first, linearity, and, second, consistency with the Lie algebra structure:

$$\psi[e_1, e_2] = [\psi(e_1), \psi(e_2)].$$

The commutator $[\cdot, \cdot]$ of two vector fields on M^n can be defined in a purely geometrical manner by means of the one-parameter groups of diffeomorphisms associated with these fields. In the local coordinates $x_1, \ldots, x_n$, the commutator of the two vector fields corresponding to the systems of ordinary differential equations $\dot{x} = F(x)$ and $\dot{x} = G(x)$ corresponds to the system with right-hand member $(\partial F/\partial x)G - (\partial G/\partial x)F$.

One of the problems which naturally arise in the theory of group actions is that of studying such actions in the neighborhood of a fixed point x_0. If we associate with each element g of G the linear part of the diffeomorphism $\varphi(\cdot, g)$ at x_0, we obtain a linear action of G on the space $TM^n|_{x_0}$. Thus, there arises the natural question of the equivalence of a nonlinear action in the neighborhood of x_0 and the linear part of this action. The answer is always affirmative in the case of a compact group G.

THEOREM (BOCHNER). *Let G be a compact group acting on a manifold M^n and having a fixed point $x_0 \in M^n$. Then there exist a neighborhood U of x_0 in M^n and a neighborhood V of the origin in $TM^n|_{x_0}$, both invariant with respect to the action of G, such that the nonlinear action of G on U is diffeomorphic to the linear action of G on V.*

The idea of Bochner's proof is very simple. We introduce some Riemannian metric on M^n. Averaging this metric with respect to the Haar measure on G, we obtain a G-invariant Riemannian metric on M^n. Hence every diffeomorphism $\varphi(\cdot, g)$ becomes an isometry, and in particular transforms geodesics into geodesics. Consider a ball V of sufficiently small radius centered at the origin in $TM^n|_{x_0}$. Clearly this ball is invariant with respect to G. Now, using the exponential mapping, we map V into M^n and denote the image of V by U. It follows from the invariance of the metric that this mapping of V onto U establishes a differentiable correspondence between the linear and nonlinear actions of G. Another proof of Bochner's theorem will be described below.

EXERCISE. Prove that the set of fixed points of any action of a compact group on a compact manifold M^n is the union of a finite number of manifolds of various dimensions.

EXERCISE. Prove that if φ_1 and φ_2 are two C^1-close actions of a compact group G in the neighborhood of a fixed point x_0, then φ_1 is differentially equivalent to φ_2.

Let $\varphi: M^n \longrightarrow M^n$ be a diffeomorphism of M^n and X a vector field on M^n. We define a vector field $\varphi_* X$ in the following manner:

$$(\varphi_* X)(x) = \varphi'\big|_{\varphi^{-1}(x)} X(\varphi^{-1}(x)).$$

LEMMA 1. *Let X be the vector field on $\mathbf{R}^n$ defined by $X = x_1\partial/\partial x_1 + \cdots + x_n\partial/\partial x_n$. Then the equality $\varphi_* X = X$ implies that the diffeomorphism φ is linear.*

PROOF. From the equality $\varphi_* X = X$ it follows that $\varphi \circ \varphi_t = \varphi_t \circ \varphi$ for any $t \in \mathbf{R}$, where $\{\varphi_t\}$ is the one-parameter group of diffeomorphisms generated by the field X. Since X is the "radial" vector field, the diffeomorphism φ_t is simply multiplication by e^t in $\mathbf{R}^n$. Hence $\varphi(x) = e^t\varphi(e^{-t}x)$. It follows from the differentiability of φ that $\varphi(x) = \varphi'|_0 \cdot x + O(|x|)$, and so

$$\varphi(x) = e^t(\varphi'|_0 \cdot e^{-t}x + O(|e^{-t}x|)) = \varphi'|_0 \cdot x + e^t \cdot O(e^{-t}|x|).$$

Letting $t \longrightarrow +\infty$, we obtain $\varphi(x) = \varphi'|_0 \cdot x$. Q.E.D.

LEMMA 2 (POINCARÉ, STERNBERG). *By a smooth change of coordinates the system $\dot{x} = x + \cdots$ can be reduced to the form $\dot{y} = y$ in some neighborhood of the origin in $\mathbf{R}^n$.*

OUTLINE OF THE PROOF. Let φ_t denote the one-parameter group of diffeomorphisms corresponding to the nonlinear system. Assume that there exists a diffeomorphism $h: V \longrightarrow \mathbf{R}^n$, $h(x) = x + \cdots$, reducing the system to linear form. Then $h \circ \varphi_t \circ e^{-t} = e^t \circ h \circ e^{-t}$ and

$$\lim_{t \to +\infty} (h \circ \varphi_t \circ e^{-t}) = h'|_0 = \text{identity}.$$

This formula suggests how one may determine h. In fact, it can be shown that as $t \longrightarrow +\infty$ the diffeomorphism $h_t = e^t \circ \varphi_{-t}$ converges, in some neighborhood of the origin, to a diffeomorphism h which conjugates the linear and nonlinear systems. In the case of a real (or complex) analytic right-hand side we obtain an analytic diffeomorphism h.

SECOND PROOF OF BOCHNER'S THEOREM. Consider a local coordinate system centered at x_0, and let $X = x_1\partial/\partial x_1 + \cdots + x_n\partial/\partial x_n$. Consider the field $\widetilde{X} = \int_G (g_* X)\,dg$ obtained from X by averaging by means of Haar measure on G. Clearly, $g_*\widetilde{X} = \widetilde{X}$ for any $g \in G$. But the field $\widetilde{X}$ has the form $\widetilde{X} = X + \cdots$, and by Lemma 2 there exist local coordinates $y_1, \ldots, y_n$ such that $\widetilde{X} = y_1\partial/\partial y_1 + \cdots + y_n\partial/\partial y_n$. By Lemma 1, the action of G is linear in these coordinates.

The following theorem may be proved in an entirely similar manner.

THEOREM (CARTAN). *Let $\varphi: G \times \mathbf{C}^n \longrightarrow \mathbf{C}^n$ be an analytic action of a compact group G on $\mathbf{C}^n$, having the origin as a fixed point. Then in some neighborhood of this fixed point one can choose analytic local coordinates in which the action of G is linear.*

The "unitary trick" enables one to prove an analogous statement for an analytic action of any semisimple Lie group.

Some results can also be obtained in the case of a nonsemisimple noncompact group G. For example, an action of the group $\mathbf{R}^k$ on $\mathbf{R}^n$ is determined by prescribing k mutually commuting vector fields on $\mathbf{R}^n$ (corresponding to the elements of some basis for the Lie algebra of the group $\mathbf{R}^k$): $[X_i, X_j] = 0$, $1 \leqslant i \leqslant j \leqslant k$. Note that any action of $\mathbf{R}$ defined by a single field X can be converted into an action of $\mathbf{R}^k$ by letting $X_1 = X$ and $X_i \equiv 0$, $i = 2, \ldots, k$. Therefore the classification of actions of $\mathbf{R}^k$ subsumes the classification of actions of $\mathbf{R}$. Apparently, by imposing certain conditions which exclude reducibility into a smaller group, e.g. requiring that any orbit be k-dimensional, one can study the action of $\mathbf{R}^k$ on manifolds of dimension $\leqslant k+1$. The theory which arises should be analogous to the theory of flows on two-dimensional surfaces. As for the case of manifolds with dimension $\geqslant k+2$, one can apparently construct examples of a type analogous to the example of Smale.

Once again, it should be noted that the space of actions of G may behave very badly. Thus, for example, the space of actions of $\mathbf{R}^2$ coincides with the space of solutions of the nonlinear partial differential equation $[X_1, X_2] = 0$ for the unknown vector fields X_1 and X_2, or, in local coordinates,

$$\frac{\partial X_1}{\partial x} \cdot X_2 - \frac{\partial X_2}{\partial x} \cdot X_1 = 0.$$

If one confines oneself to linear actions of $\mathbf{R}^2$ (or $\mathbf{R}^k$), the corresponding equations become nonlinear algebraic equations, having solutions which also constitute quite a complicated structure. In the case of $\mathbf{R}^2$, this system of algebraic equations has the form $AB - BA = 0$, where A and B are square matrices of order n. Thus, although in some cases we are able to linearize the action in a neighborhood of a fixed point, this cannot be regarded as the final solution to the problem.

The following simple theorem concerning the linearization of an action of $\mathbf{R}^k$ is formulated and proved, using entirely different notation, in the work of Perov on multidimensional differential equations.

THEOREM. *Let φ be an action of $\mathbf{R}^k$ on $\mathbf{R}^n$, having the origin as a fixed point. Suppose that the field $X = x_1\partial/\partial x_1 + x_2\partial/x_2 + \cdots$ is among the vector fields corresponding to the elements of the Lie algebra of $\mathbf{R}^k$. Then there is a local coordinate system, with center at* 0, *in which the action is linear.*

PROOF. From the commutativity of the group $G = \mathbf{R}^k$ it follows that $g_*X = X$ for any $g \in \mathbf{R}^k$. By Lemma 2, there exists a coordinate system $y_1, \ldots, y_n$ in which the field X takes the form $X = y_1\partial/\partial y_1 + \cdots + y_n\partial/\partial y_n$. By Lemma 1, the action is linear in this very same coordinate system.

Actions of compact groups, as might be expected, also possess many remarkable properties. We have already seen that the action of a compact group in the neighborhood of a fixed point is equivalent to a linear action, and the set of mutually inequivalent linear representations of a fixed compact group in a space of fixed dimension is finite. Hence the action of a compact group in the neighborhood of a fixed point is differentiably stable. Since the action of a compact group in the neighborhood of a prescribed orbit is completely determined by the action of the stationary subgroup of some point of this orbit, the action of a compact group in the neighborhood of an orbit of a fixed type is also differentially stable. In fact, a more general assertion is valid.

THEOREM (PALAIS). *Let φ be a C^∞-action of a compact group G on a compact manifold M^n. Then any action which is sufficiently close to φ in the C^∞ topology is differentiably equivalent to φ.*

Thus the set of pairwise inequivalent actions of a compact group G on a fixed manifold M^n is at most countable. However, the questions of existence of nontrivial actions and the classification of such actions constitute complex topological problems.

BIBLIOGRAPHY

1. A. Andronov and L. Pontrjagin, *Systèmes grossiers,* C. R. (Dokl.) Acad. Sci. URSS **14** (1937), 247–250.

2. M. M. Peixoto, *Structural stability on two-dimensional manifolds,* Topology **1** (1962), 101–120.

3. ———, *Structural stability on two-dimensional manifolds. A further remark,* Topology **2** (1963), 179–180.

4. S. Smale, *Structurally stable systems are not dense,* Amer. J. Math. **88** (1966), 491–496.

5. R. F. Williams, *The "DA" maps of Smale and structural stability,* Global Analysis, Proc. Sympos. Pure Math., vol. 14, Amer. Math. Soc. Providence, R. I., 1970, pp. 329–334.

6. D. M. Grobman, *Topological classification of neighborhoods of a singularity in n-space,* Mat. Sb. **56 (98)** (1962), 77–94. (Russian)

7. Philip Hartman, *On the local linearization of differential equations,* Proc. Amer. Math. Soc. **14** (1963), 568–573.

8. S. Smale, *Dynamical systems and the topological conjugacy problem for diffeomorphisms,* Proc. Internat. Congr. Math. (Stockholm, 1962), Inst. Mittag-Leffler, Djursholm, 1963, pp. 490–496.

9. R. Thom and H. Levine, *Singularities of differentiable maps,* Lecture notes, Univ. Bonn, 1959.

10. Charles C. Pugh, *An improved closing lemma and a general density theorem,* Amer. J. Math. **89** (1967), 1010–1021.

11. ———, *The closing lemma,* Amer. J. Math. **89** (1967), 956–1009.

12. J. Palis and S. Smale, *Structural stability theorems,* Global Analysis, Proc. Sympos. Pure Math., vol. 14, Amer. Math. Soc., Providence, R. I., 1970, pp. 223–231.

13. D. V. Anosov, *Geodesic flows on closed Riemannian manifolds with negative curvature,* Trudy Mat. Inst. Steklov. **90** (1967); English transl., Proc. Steklov Inst. Math. **90** (1967) (1969).

Translated by E. J. BRODY

Amer. Math. Soc. Transl.
(2) Vol. 116, 1981

DYNAMICAL SYSTEMS WITH HYPERBOLIC STRUCTURE*

A. B. KATOK

CONTENTS

Introduction

The lectures to be presented here represent an attempt to give an account of a certain "connected component" of the theory of smooth dynamical systems. The central idea of our considerations is that of a hyperbolic set (a closed set, invariant with respect to some diffeomorphism, in a neighborhood of which the diffeomorphism is contractive along some directions and expansive along the others). Hyperbolic sets possess strong stability properties. For example, they are preserved by small perturbations of the diffeomorphism, and in this connection it is quite remarkable that the topological structure of the hyperbolic set itself may be very complicated (see Examples 1.5 and 1.6 of §1, below). On the other hand, the action of a diffeomorphism on a sufficiently large hyperbolic set bears little similarity to those situations which are traditionally considered in the qualitative theory of differential equations (conditionally periodic motions and their various generalizations), and has, rather a "stochastic" character (see, in particular, Example 1.5, where one may obtain a sequence of independent random variables by introducing a natural invariant measure on the hyperbolic set). Perhaps the most important reason leading to the consideration of hyperbolic sets is the fact that this notion covers situations arising in many problems of classical (in particular, celestial) mechanics, differential geometry, etc. (see Part III of the lectures of V. M. Alekseev contained in this volume).

In these lectures, we shall confine ourselves to diffeomorphisms; that is, to the case of dynamical systems with discrete time. Most of the definitions and results

---

1980 *Mathematics Subject Classification.* Primary 58F15; Secondary 58F20.

* Translation of Ninth Math. Summer School (Kaciveli, 1971), Izdanie Inst. Mat. Akad. Nauk Ukrain. SSR, Kiev, 1972, pp. 125–211. MR 51 #14160.

discussed below can be modified and applied to flows (that is, dynamical systems with continuous time (see [1] and [2])) and, in part, to smooth actions of more general Lie groups. We shall not do this, in order to avoid encumbering the presentation with parallel formulations. The examples of diffeomorphisms given in these lectures may be converted into examples of flows by means of a construction described in §2 of the lecture notes of A. G. Kušnirenko, also contained in this volume. However, the converse is not true, since the construction of a global succession function is not always possible for flows. For example, far more Anosov flows are known than Anosov diffeomorphisms, since, in addition to the suspensions of Anosov diffeomorphisms, there also exist geodesic flows on manifolds of negative curvature and certain other dynamical systems (see [1], §2). The existence of an Anosov flow imposes less stringent restrictions upon the structure of a manifold than the existence of an Anosov diffeomorphism, and therefore it appears that the classification of Anosov flows should turn out to be substantially richer than the classification of Anosov diffeomorphisms (see §4 below).

The various parts of these lectures are not homogeneous as regards the character of the presentation. §1, §2, the first half of §4, and §5 are presented in greatest detail. The proofs given there are sufficiently explicit, although we nevertheless omit certain details. In §2 we prove Anosov's theorem on families of ϵ-trajectories, which may be regarded as one of the two basic technical results relating to hyperbolic sets. We shall obtain all our assertions concerning the stability of hyperbolic sets (in particular, structural stability of Anosov diffeomorphisms) as consequences of this theorem. The second of these two most important results, i.e. the theorem on stable manifolds, is formulated in §3. On the whole, §§1, 2, 3, subsections 1 and 2 of §4, and §5 contain a sufficiently coherent presentation of the foundations of the theory of hyperbolic sets; the only serious gap in the presentation is the absence of a proof of the theorem on stable manifolds. The second half of §4 and §6 are of a survey character; we have attempted to give major consideration to the new results obtained since the appearance of Smale's survey article [2]. §7 contains a number of definitions, examples and assertions which appear to be somewhat disconnected, but, which in the author's opinion, should in due time become parts of some general picture. In the Appendix we give a proof of one known result concerning the asymptotics of the number of periodic trajectories, which once again demonstrates the usefulness of the theorem on families of ϵ-trajectories.

The text contains a series of exercises. Some of these merely require correct application of the definitions, or repetition of the preceding arguments, while others require the application of certain additional (but, as a rule, fairly simple) considerations. In addition, some of the assertions which are stated without proof may also be regarded as exercises (Lemmas 3.1–3.5, Theorem 4.1 and its generalization to locally maximal hyperbolic sets (§5.3), the assertions of Examples 5.3, 6.1, 6.2, and Propositions 7.1 and 7.2). Finally, the formulation and proof, for the case of flows, of the statements which correspond to the basic assertions proved for diffeomorphisms in these lectures also constitute a series of very useful exercises. In this connection, one should, of course, take account of the remark in §3 of Kušnirenko's lectures concerning the difference between the notions of stability for flows and diffeomorphisms, since transformations of the time scale must be disregarded.

The connection of the present lectures with Kušnirenko's is loose, and largely reduces to the references concerning certain definitions and (to a lesser extent) examples from those lectures (also, in Kušnirenko's lectures certain arguments are presented, on the basis of which one can prove the theorem on stable manifolds stated in §3). Nevertheless, the unprepared reader should by all means familiarize himself with Kušnirenko's lectures before reading ours. The point is that the character of our presentation is considerably more formal than that of Kušnirenko's and we are not especially concerned about clarifying the relation of our matters under consideration with the more traditional questions of the theory of differential equations.

We assume that the reader is familiar with the elementary facts concerning smooth manifolds, tangent bundles, differentials of mappings, Riemannian metrics, and exponential mappings. All this information may be found in the monographs of Lang [10] (Chapters 1, 2, 3 and 7), Bishop and Crittenden [39] (Chapters 1, 3 and 7) or Sternberg [40] (Chapter 2 and §6 of Chapter 4). However, we actually make use of little more than some definitions. Also, it is helpful to be acquainted with the elementary concepts of the topological theory of dynamical systems, for example, to the extent of Chapter 4 of Nemyckiĭ and Stepanov's book [41].

§1. Hyperbolic sets

1. Let M be a smooth manifold, U an open submanifold of M, and $f\colon U \longrightarrow M$ a diffeomorphism. It is simplest to merely assume that M and f are of smoothness class C^∞; however, for the validity of most of the results, it is sufficient for f to be of class C^1, and only in certain cases is greater smoothness required. In the definition of a hyperbolic set given below, use is made of a Riemannian metric on M, which may be chosen arbitrarily, since this definition does not depend upon the choice of this metric (see Proposition 1.1). The subset $\Lambda \subset M$ is said to be *invariant* with respect to f if Λ belongs to the domain of f^{-1} and $x \in \Lambda$ implies that $f(x), f^{-1}(x) \in \Lambda$.

The differential Df acts on the tangent bundle; that is, $Df\colon\ TM \longrightarrow TM$, or, pointwise, $Df_x\colon\ T_xM \longrightarrow T_{f(x)}M$, and this mapping is an isomorphism of the respective linear spaces. A Riemannian metric on the manifold defines an inner product (and hence a norm) on each tangent space $T_x M$.

DEFINITION 1.1. A compact subset $\Lambda \subset M$, invariant with respect to the diffeomorphism $f\colon\ U \longrightarrow M$, is said to be *hyperbolic* if for each point $x \in \Lambda$ there exists a pair of linear subspaces E^s_x and E^u_x of the tangent space T_xM, such that:

1.1.1. $T_xM = E^s_x \oplus E^u_x$ (direct, but not necessarily orthogonal, sum), $\dim E^s_x = k(x)$, $\dim E^u_x = l(x)$, $k(x) + l(x) = n$, with $k(x)$ and $l(x)$ depending continuously upon x;

1.1.2. $Df_x(E^s_x) = E^s_{f(x)}$ and $Df_x E^u_x = E^u_{f(x)}$;

1.1.3. there exist constants $c > 0$ and $0 < \lambda < 1$ such that

$$\|Df_x^n u\| \leqslant c\lambda^n \|u\| \tag{1.1}$$

and

$$\|Df_x^{-n} v\| \leqslant c\lambda^n \|v\|, \tag{1.2}$$

where $u \in E_x^s$, $v \in E_x^u$, and $n > 0$.

We write $\bigcup_{x \in \Lambda} E_x^s = E^s$ and $\bigcup_{x \in \Lambda} E_x^u = E^u$.

2. Let us consider some simple consequences of Definition 1.1.

PROPOSITION 1.1. *If the set Λ is hyperbolic with respect to some Riemannian metric on M, then it is also hyperbolic with respect to any other Riemannian metric; moreover, the constant λ in condition* 1.1.3 *may be chosen independently of the metric.*

PROOF. Let $\|\cdot\|'$ denote the norm induced by the second metric. By the compactness of Λ, one can find positive constants k and K such that $k \leqslant \|w\|'/\|w\| < K$ for $w \in T_\Lambda M$. Thus, for any $u \in E_x^s$ and $v \in E_x^u$,

$$\|Df_x^n u\|' \leqslant K \|Df_x^n u\| \leqslant K c \lambda^n \|u\| \leqslant K k^{-1} c \lambda^n \|u\|',$$

$$\|Df_x^{-n} v\| \leqslant K \|Df_x^{-n} v\| \leqslant K c \lambda^n \|v\| \leqslant K k^{-1} c \lambda^n \|v\|'.$$

PROPOSITION 1.2. *The subspaces E_x^s and E_x^u depend continuously upon the point $x \in \Lambda$.*

PROOF. First, observe that E_x^s consists of all the vectors $u \in T_x \Lambda$ for which the inequalities (1.1) are satisfied. In fact, if $w \in T_x \Lambda$, $w = u + v$, $u \in E_x^s$, $v \in E_x^u$, then

$$\|Df_x^n w\| \geqslant \|Df_x^n v\| - \|Df_x^n u\| \geqslant c^{-1} \lambda^{-n} \|v\| - c \lambda^n \|u\| \longrightarrow \infty \text{ for } v \neq 0.$$

The set $E^{s,n}$ of vectors $u \in T_\Lambda M$ satisfying (1.1) for a given n is obviously closed, and therefore the set $E^s = \bigcap_{n \geqslant 0} E^{s,n}$ is also closed. Since the dimension of E_x^s is locally constant, the proposition is proved.

PROPOSITION 1.3. *For any hyperbolic set Λ there exists a Riemannian metric on M such that for some μ, $0 < \mu < 1$,*

$$\|Dfu\| \leqslant \mu \|u\|, \quad \|Df^{-1} v\| \leqslant \mu \|v\|, \tag{1.3}$$

where $u \in E^s$ and $v \in E^u$.

PROOF. Given an arbitrary Riemannian metric, we denote the inner product of the vectors $u, v \in T_x M$ by $L(u, v)$. Let

$$(f^* L)(u, v) = L(Df_x u, Df_x v).$$

We define a new inner product on $T_\Lambda M$ as follows:

$$L^*(u_1 + v_1, u_2 + v_2) = \sum_{n \geqslant 0} f^{n*}L(u_1, u_2) + f^{-n*}L(v_1, v_2), \tag{1.4}$$

where $u_1, u_2 \in E^s_x$, $v_1, v_2 \in E^u_x$ and $x \in \Lambda$. For $u_1 = u_2 = u$ and $v_1 = v_2 = v$ we have, by virtue of (1.1) and (1.2),

$$f^{n*}L(u, u) + f^{-n*}L(v, v) = \|Df^n u\|^2 + \|Df^{-n}v\|^2$$
$$\leqslant c^2\lambda^{2n}(\|u\|^2 + \|v\|^2);$$

hence (1.4) converges.

Let $u \in E^s$. Using the notation $L^*(w, w)^{1/2} = \|u\|^*$, we then have

$$(\|u\|^*)^2 = \|u\|^2 + (\|Dfu\|^*)^2,$$

$$\|Dfu\|^* = \left(\sum_{n=1}^{\infty} \|Df^n u\|^2\right)^{1/2} \leqslant \left(\sum_{n=1}^{\infty} c^2\lambda^{2n}\|u\|^2\right)^{1/2} = \frac{c\lambda}{1-\lambda}\|u\|.$$

Therefore

$$\|Dfu\|^* \leqslant \sqrt{1 - \left(\frac{1-\lambda}{c\lambda}\right)^2}\,\|u\|^*.$$

Hence, using a similar computation for $v \in E^u$, we obtain

$$\|Dfu\|^* < \sqrt{1 - \left(\frac{1-\lambda}{2c\lambda}\right)^2}\,\|u\|^*,$$
$$\|Df^{-1}v\|^* < \sqrt{1 - \left(\frac{1-\lambda}{2c\lambda}\right)^2}\,\|v\|^*. \tag{1.5}$$

We extend L^* to a continuous positive definite inner product on TM and approximate this product by an inner product L_1 of class C^∞ such that the inequalities (1.5) remain valid for the norm induced by L_1.

DEFINITION 1.2. A Riemannian metric which satisfies inequalities (1.3) is said to be *Ljapunov* (see [1], §7).(1)

3. Examples of hyperbolic sets.

EXAMPLE 1.1. The simplest example appears when the set Λ consists of a single point p. Then p is a fixed point of the diffeomorphism f, Df is a linear operator of T_pM, and conditions (1.1) and (1.2) mean that p is a hyperbolic fixed point (see §5.1 in Kušnirenko's lectures, contained in this volume); that is, the moduli of all the eigenvalues of the operator $Df(p)$ are different from 1. In other words, $(\operatorname{spec} Df(p)) \cap \Delta = \varnothing$, where Δ is the unit circle in the complex plane. (Later on, we shall show that the definition of a hyperbolic set can also be similarly formulated in the general case.)

If $\Lambda = \{p, \ldots, f^{n-1}(p)\}$ is a periodic trajectory with period n, then the hyperbolicity of Λ is equivalent to the condition $(\operatorname{spec}(Df^n_p)) \cap \Delta = \varnothing$.

(1) We remark that Smale ([2], p. 759) attributes the construction of such a metric to Mather.

DEFINITION 1.3. If the entire manifold M is a hyperbolic set of the diffeomorphism $f\colon M \longrightarrow M$, then f is called an *Anosov diffeomorphism.* *

EXAMPLE 1.2. As an elementary example of an Anosov diffeomorphism, one may cite the diffeomorphism f_A of the two-dimensional torus $\mathbf{T}^2$ induced by the matrix $A = \begin{pmatrix} 2 & 1 \\ 1 & 1 \end{pmatrix}$, which is considered in some detail in Kušnirenko's lectures (Example 12 of §8). In this case the subspace E_x^s consists of the vectors parallel to the vector e_1 (the eigenvector of the matrix A with the smaller eigenvalue λ); the subspace E_x^u consists of the vectors parallel to the second eigenvector e_2.

Any metric on $\mathbf{T}^2$ which is invariant with respect to the group action is Ljapunov.

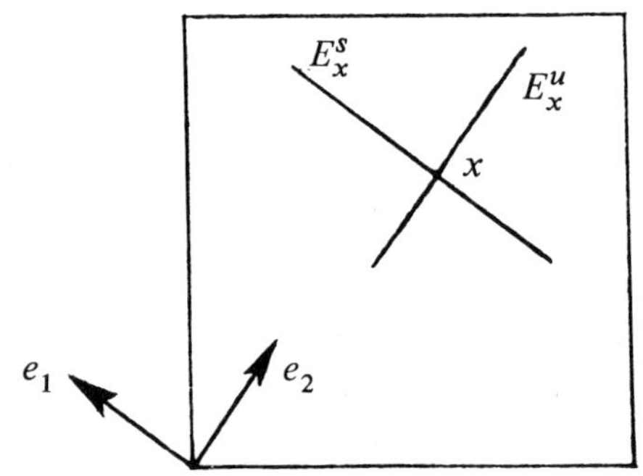

FIGURE 1

This example admits a natural generalization:

EXAMPLE 1.3. Let $A \in SL(n, \mathbf{Z})$, $(\operatorname{spec} A) \cap \Delta = \varnothing$. Then the matrix A induces an automorphism of the n-dimensional torus $\mathbf{T}^n = \mathbf{R}^n/\mathbf{Z}^n$, which constitutes an Anosov diffeomorphism.

Let $\operatorname{spec} A = S_1 \cup S_2$, where S_1 consists of the eigenvalues with modulus less than 1, and S_2 consists of the eigenvalues with modulus greater than 1.

The partition of the spectrum of A into S_1 and S_2 corresponds to a decomposition $\mathbf{R}^n = E_0^s \oplus E_0^u$. By parallel translation of this decomposition into each point $x \in \mathbf{T}^n$, we obtain a decomposition $E_x^s \oplus E_x^u$, satisfying the conditions of Definition 1.1 with

$$\lambda = \max\left(\max_{\mu \in S_1} |\mu|,\ \max_{\mu \in S_2} |\mu^{-1}|\right).$$

A Ljapunov Riemannian metric on $\mathbf{T}^n$ is induced by any scalar product on $\mathbf{R}^n$ with respect to which $\|A|_{E_0^s}\| < 1$ and $\|A|_{E_0^u}\| < 1$.

**Editor's note.* This translation uses the Western terminology. In his original definition, Anosov called them (U)-diffeomorphisms (see [1]).

EXAMPLE 1.4. Further examples of Anosov diffeomorphisms may be constructed by using a nilpotent simply-connected Lie group N in place of the commutative Lie group $\mathbf{R}^n$. If Γ is a discrete subgroup of N, if the space N/Γ is compact, and if f is an automorphism of N such that $(\operatorname{spec} Df_e) \cap \Delta = \varnothing$ (where e is the identity element of the group) and $f(\Gamma) = \Gamma$, then f determines a diffeomorphism $N/\Gamma \longrightarrow N/\Gamma$, which again constitutes an Anosov diffeomorphism. Concrete examples of this type were first given by Smale ([2], §I.3); a still more general construction of Anosov diffeomorphisms is described in [3].

All these examples are of a specifically algebraic nature. However, it turns out that any diffeomorphism which is sufficiently close to an Anosov diffeomorphism in the C^1-topology is also an Anosov diffeomorphism (see Corollary 2.1 below).

EXAMPLE 1.5. We now proceed to construct a hyperbolic set of a completely different topological character. Consider a diffeomorphism f of the square Δ onto the "horseshoe" $f(\Delta)$, linear on the inverse images of the two rectangles Δ_0 and Δ_1 which constitute the intersection of Δ and $f(\Delta)$ (see Figure 2). The intersection $\Delta \cap f(\Delta) \cap f^2(\Delta)$ consists of four rectangles $\Delta_{ij} = \Delta_i \cap f(\Delta_j)$, $i, j = 0, 1$. Considering the intersections of the successive images of Δ, we find that $\Lambda^+ = \bigcap_0^{+\infty} f^n(\Delta)$ represents the direct product of a Cantor perfect set and a horizontal segment. Similarly, $\Lambda^- = \bigcap_0^{+\infty} f^{-n}(\Delta)$ is the direct product of a Cantor perfect set and a vertical segment.

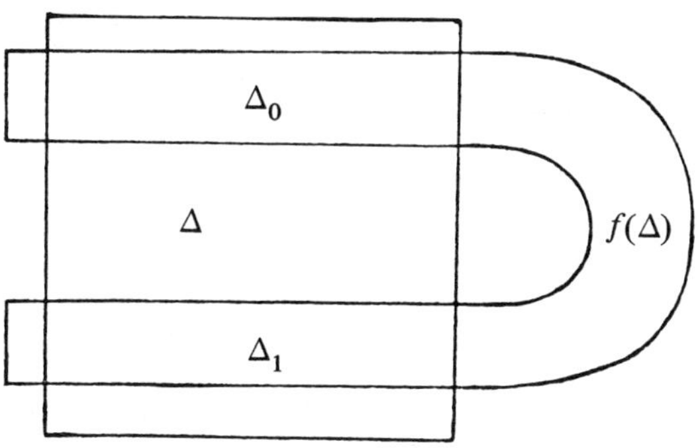

FIGURE 2

The maximal subset of Δ which is invariant with respect to f, i.e.

$$\Lambda = \Lambda^+ \cap \Lambda^- = \bigcap_{n=-\infty}^{+\infty} f^n(\Delta),$$

will be the product of two Cantor perfect sets. The hyperbolicity of Λ is clear, since f is linear on the inverse images of $\Delta \cap f(\Delta)$. Hence, for $x \in \Lambda$, Df_x is a fixed linear mapping which is contractive in the vertical direction and expansive in the horizontal direction; E_x^s and E_x^u are the vertical and horizontal lines, respectively, passing through x. Let us investigate the diffeomorphism f on the set Λ. Every horizontal segment of Λ^+ may be represented in the form of an intersection $\Delta_{i_0} \cap f(\Delta_{i_1}) \cap \cdots \cap f^n(\Delta_{i_n}) \cap \cdots$, where $i_k = 0, 1$. We denote this intersection by $\Delta_{i_0 i_1 \cdots}$. Similarly, every vertical segment of Λ^- may be represented in the form

$$\Delta^{i_{-1}i_{-2}\cdots i_{-n}\cdots} = \bigcap_{n=1}^{\infty} f^{-n}(\Delta_{i_n}).$$

Every point of Λ is the intersection of a vertical and a horizontal segment, and with this one may naturally associate a doubly infinite sequence of zeros and ones:

$$\omega = \{\cdots i_{-2}i_{-1}i_0i_1i_2\cdots\},$$

$$x_\omega = \Delta^{i_{-1}i_{-2}\cdots i_{-n}\cdots} \cap \Delta_{i_0i_1\cdots i_n\cdots} = \bigcap_{n=-\infty}^{+\infty} f^n(\Delta_{i_n}).$$

Conversely, any doubly infinite sequence of zeros and ones determines a certain point $x_\omega \in \Lambda$. We define a mapping S in the space Ω of doubly infinite sequences of zeros and ones by setting

$$x_{S_\omega} = f(x_\omega) = \bigcap f^{n+1}(\Delta_{i_n}) = \bigcap f^n(\Delta_{i_{n-1}}).$$

Clearly, $i_n(S\omega) = i_{n-1}(\omega)$. Hence S is just a shift by one place to the right.

Define the mapping φ: $\Omega \longrightarrow \Lambda$ by $\varphi(\omega) = x_\omega$. Thus $\varphi S = f\varphi$. This representation makes it possible to find the periodic points of $f|_\Lambda$ and prove their density in Λ.

One may introduce a topology in the space Ω by regarding it as the Tychonoff product of two-point spaces. The cylinders (i.e., sets consisting of the elements ω such that $i_n(\omega)$ is fixed for a finite number of places and unrestricted otherwise) form a basis for the open sets. These cylinders are closed, since the complement of a cylinder is the union of a finite number of cylinders. It is not difficult to verify that this topology coincides with the topology of Λ induced by $\mathbf{R}^2$.

PROPOSITION 1.4. *The periodic points of the diffeomorphism f are dense in Λ.*

PROOF. Let $x_\omega \in \Lambda$. Given any neighborhood U of x_ω, we shall prove that U contains a periodic point of f. Let $\varphi^{-1}(U) = V$. There exists an $N > 0$ such that $C = \{\omega'\colon\ i_n(\omega') = i_n(\omega), |n| \leqslant N\} \subset V(\omega)$. Take a point $\omega_N \in C$ for which i_n is periodically repeated beyond the end-points of the interval $-N < n \leqslant N$. Clearly, $S^{2N+1}\omega_N = \omega_N$, and therefore $f^{2N+1}x_{\omega_N} = x_{\omega_N}$; that is, $x_{\omega_N} \in U(x_\omega)$ is a periodic point.

EXERCISE 1.1. Show that the homeomorphism S is topologically transitive.

One can introduce a group structure in the space Ω, taking coordinatewise addition modulo 2 as the group operation; thus $\Omega = \Pi_{-\infty}^{+\infty} \mathbf{Z}_2$. The homeomorphism S then becomes a group automorphism, like the Anosov diffeomorphisms considered in Examples 1.2 and 1.3. We shall describe another example arising in a similar situation.

EXAMPLE 1.6. The construction of many examples of hyperbolic sets, affording insight into very subtle effects, makes use of a device known as the "inverse limit of a mapping." This device permits the construction of a diffeomorphism $f\colon \mathbf{D}^{n+1} \times M^n \longrightarrow \mathbf{D}^{n+1} \times M^n$ with a hyperbolic set Λ for any given smooth (not one-to-one) mapping $g\colon M^n \longrightarrow M^n$ with repulsive set Ω. A general description of this construction is given in Smale's survey article [2] (p. 788).* We shall confine ourselves to the simplest example of this construction, in which $M^n = S^1$, $g\colon S^1 \longrightarrow S^1$ and $g(\varphi) = 2\varphi \pmod 1$; that is, g is a uniform dilation of multiplicity of the circle.

Consider the immersion $\psi_0\colon S^1 \longrightarrow M = \mathbf{D}^2 \times S^1$, $\psi_0(\varphi) = (0, 0, 2\varphi)$. This immersion can be approximated in the C^1-topology by an imbedding of the form $\psi(\varphi) = (\psi_1(\varphi), \psi_2(\varphi), 2\varphi)$, and then ψ can be extended to a mapping $f\colon M \longrightarrow M$ such that every section $\mathbf{D}^2 \times \varphi_0$ of the solid torus $\mathbf{D}^2 \times S^1$ is mapped into the section $\mathbf{D}^2 \times 2\varphi_0$, and, moreover, is uniformly contracted. Here is a concrete example:

$$f(x, y, \varphi) = (10^{-10}x + \tfrac{1}{2}\cos\varphi,\ 10^{-10}y + \tfrac{1}{2}\sin\varphi,\ 2\varphi).$$

The intersection $\Lambda = M \cap f(M) \cap \cdots$ is a solenoid. Topologically, this is a locally trivial fibration with S^1 as fiber and the Cantor perfect set as base. The solenoid Λ is connected, but not locally connected.

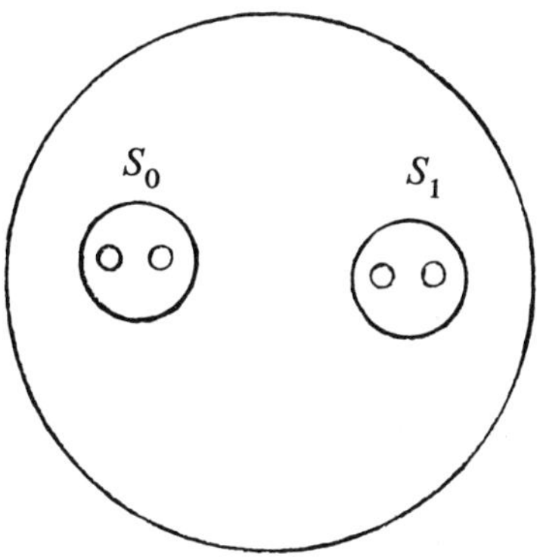

FIGURE 3

In fact, let us fix $\varphi_0 \in S^1$ and consider the intersection of the section $S = \mathbf{D}^2 \times \varphi_0$ with the successive images $f^n(M)$ (see Figure 3). $S \cap fM$ is the union of two disjoint disks S_1 and S_2; one of these represents the image of the section $\mathbf{D}^2 \times \varphi_0/2$, and the other is the image of the section $\mathbf{D}^2 \times (\varphi_0/2 + \tfrac{1}{2})$. The intersection $f^2M \cap S$ is the union of the four disks

$$S_{00} = f^2(\mathbf{D}^2 \times \varphi_0/4), \qquad S_{01} = f^2(\mathbf{D}^2 \times (\varphi_0/4 + \tfrac{1}{2})),$$
$$S_{10} = f^2(\mathbf{D}^2 \times (\varphi_0/4 + \tfrac{1}{4})), \qquad S_{11} = f^2(\mathbf{D}^2 \times (\varphi_0/4 + \tfrac{3}{4})).$$

---

* *Editor's note.* With reference to the sentence immediately preceding the figure on p. 788 of [2], V. M. Alekseev and E. V. Maĭkov, the translators of the Russian version of [2], remark as follows.

S' is interpreted as the set of complex z such that $|z| = 1$. Taking $M = S^2$, the Riemann sphere, and defining the endomorphism f by $f(z) = \tfrac{1}{2}(3z - z^3)$ (this mapping is studied in detail in [50]; it is nonexpanding), we obtain (J. Guckenheimer) an example of a basic set that at no point has the local structure of the product of a manifold and a Cantor set.

Clearly, the ensuing process leads to a Cantor set, each point λ of which may be uniquely defined by specifying the coordinate φ of all its inverse images.

There exists a one-to-one correspondence between the points of the solenoid and the set Ξ of numerical sequences

$$(\xi_0, \xi_1, \ldots), \quad \text{where} \quad |\xi_0| = 1 \quad \text{and} \quad \xi_{k+1}^2 = \xi_k, k = 0, 1, \ldots. \tag{1.6}$$

In fact, let $\lambda \in \Lambda$; then we set

$$\xi(\lambda) = (\xi_0(\lambda), \xi_1(\lambda), \ldots). \tag{1.7}$$

Conversely, if $\xi_k = e^{2\pi i \varphi_k}$, then the condition $\varphi_k = \varphi(f^{-k}\lambda)$ uniquely determines a point $\lambda \in \Lambda$. In the space of all sequences of the form (1.6) one may introduce a topology by taking, as a basic neighborhood of the sequence $\xi = (\xi_0, \xi_1, \ldots)$, the set of all sequences for which the first N coordinates differ from the corresponding coordinates of the sequence ξ by no more than a prescribed number $\epsilon > 0$. Then (1.7) defines a homeomorphism $\Lambda \longrightarrow \Xi$, with respect to which the mapping $f\colon \Lambda \longrightarrow \Lambda$ is conjugate with the homeomorphism

$$a\colon \Xi \longrightarrow \Xi, \quad a(\xi_0, \xi_1, \ldots) = (\xi_0^2, \xi_0, \xi_1, \ldots) = (\xi_0^2, \xi_1^2, \xi_2^2, \ldots).$$

EXERCISE 1.2. The periodic points of the homeomorphism a are dense in Ξ, and a is topologically transitive.

The space Ξ admits the structure of a compact topological group (multiplication being coordinatewise; this is, in fact, the character group of the additive group of dyadic rationals with the discrete topology). Clearly a is a group automorphism, and consequently the diffeomorphism f which we have constructed is also of an algebraic nature with respect to the solenoid Λ. However, undue importance should not be attached to this property. Automorphisms of compact commutative groups serve as a good—but not by any means universal—model, even for "good" hyperbolic sets. For more details regarding this point, see Part 2 of V. M. Alekseev's lectures contained in this volume.

EXAMPLE 1.7. Let x_0 be a hyperbolic fixed point of the diffeomorphism $f\colon U \longrightarrow M$. Assume that the stable manifold $W^s(x_0)$ and the unstable manifold $W^u(x_0)$ (see §6 in Kušnirenko's lectures) intersect transversally in some point $x \in U$. Such a point is said to be *homoclinic*, and clearly $f^n x \longrightarrow x_0$ as $n \longrightarrow +\infty$ and also as $n \longrightarrow -\infty$. Consider the closed set $\Lambda = x_0 \cup (\bigcup_{-\infty}^{+\infty} f^n x)$; this set is invariant with respect to f. Let $\lambda \in M$, and set

$$T_\lambda M = T_\lambda(W^s(x_0)) \oplus T_\lambda(W^u(x_0)).$$

It is clear that conditions 1.1.1–1.1.3 are satisfied for this decomposition of $T_\lambda M$. Thus, Λ is a hyperbolic set. Homoclinic points and similar situations (homoclinic chains) were already known to Poincaré. They are very important, since they arise in many classical problems of mechanics, for example the n-body problem (the fact that one must deal with flows in these problems does not essentially change the situation). Smale has shown ([4] or [2], § I.5; historical information is also cited there) that the hyperbolic sets described in Example 1.5 appear in a neighborhood of a homoclinic point for some power of the diffeomorphism f. We shall not dwell upon the details of these questions, since homoclinic points are studied at length in the lectures of V. M. Alekseev, where an appropriate device (topological Markov chains) for the analysis of the hyperbolic sets arising in this manner is also described.

EXERCISE 1.3. Suppose that $W^s(x_0)$ intersects $W^u(x_0)$ in the point x, but not transversally. Then x cannot belong to any hyperbolic set.

It is easily verified that all the points of the hyperbolic sets constructed in Examples 1.1–1.7 are nonwandering. However, this is not always the case.

EXAMPLE 1.8. Let x_1 and x_2 be hyperbolic fixed points of the diffeomorphism $f\colon U \longrightarrow M$, where $\dim W^s(x_1) = \dim W^s(x_2)$. Suppose that $W^s(x_1)$ and $W^u(x_2)$ intersect transversally in the point x (which, in this case, is said to be *heteroclinic*).

The closed invariant set $\Lambda = x_1 \cup x_2 \cup (\bigcup_{-\infty}^{+\infty} f^n x)$ is hyperbolic, since the appropriate decompositions of the tangent space at x_1 and x_2 have been already given, and for $y \in \bigcup_{-\infty}^{\infty} f^n x$ we set $E_y^s = T_y W^s(x_1)$ and $E_y^u = T_y W^u(x_2)$. Clearly, $\Omega(f|_\Lambda) = x_1 \cup x_2$. Furthermore, it is a simple matter to construct an example of a diffeomorphism f of a closed manifold (e.g., the two-dimensional torus $\mathbf{T}^2$) with a heteroclinic point which is wandering on the entire manifold with respect to f (see §6, Example 6.2).

EXERCISE 1.4. Suppose that either $\dim W^s(x_1) \neq \dim W^u(x_2)$, or that the intersection of $W^s(x_1)$ and $W^u(x_2)$ is not transversal at x. Then x cannot belong to any hyperbolic set.

4. An equivalent definition of a hyperbolic set. In Examples 1.1–1.4 considered above, the hyperbolicity of the invariant set was connected with the fact that the spectrum of a certain linear operator does not intersect the unit circle (i.e., according to the usual terminology, the operator is hyperbolic). Mather [5] has observed that the general definition of an Anosov diffeomorphism can be formulated in terms of the hyperbolicity of a certain linear operator in an infinite-dimensional Banach space, and such an alternative definition turns out to be useful, for example, in proving the structural stability of an Anosov diffeomorphism. We shall prove the equivalence of Mather's definition and Definition 1.1 in the case of an arbitrary hyperbolic set. We shall make direct use of this definition only once (in the proof of Theorem 2.1 on the roughness of the property of hyperbolicity). In §2 we shall deduce the structural stability of Anosov diffeomorphisms from the theorem on families of ϵ-trajectories. However, at the basis of the proof of this latter theorem, which will play a substantial role in our presentation, there essentially lies a rather complicated variant of Mather's construction. Moreover, the usefulness of this construction has been demonstrated by recent work on structural stability (Robbin [6]) and Ω-stability (Franks [7]). We shall give further details of these results at the end of §6.

Let Λ be a closed invariant set of the diffeomorphisms $f\colon\ U \longrightarrow M$. Let $\Gamma(T_\Lambda M)$ denote the Banach space of continuous cross-sections of the bundle $T_\Lambda(M)$; that is, restrictions to Λ of continuous vector fields on M, with norm $\|u\| = \sup_{x\in\Lambda}\|u(x)\|$. The diffeomorphism f induces a linear operator

$$f_*\colon \Gamma(T_\Lambda M) \longrightarrow \Gamma(T_\Lambda M),$$
$$(f_*u)\,(x) = Dfu\,(f^{-1}(x)).$$

EXERCISE 1.5. Find the spectrum of f_* when Λ is a periodic trajectory, with period n.

THEOREM 1.1. *The set Λ is hyperbolic if and only if*

$$(\operatorname{spec} f_*) \cap \Delta = \varnothing.$$

PROOF. *Necessity.* Assume that Λ is a hyperbolic set. Then $T_\Lambda M = E^s \oplus E^u$. Clearly, $\Gamma(T_\Lambda M) = \Gamma(E^s) \oplus \Gamma(E^u)$, and, moreover, the subspaces $\Gamma(E^s)$ and $\Gamma(E^u)$ are invariant with respect to f_*.

If the metric on M is Ljapunov, then $\|f_*|_{\Gamma(E^s)}\| \leqslant \mu < 1$, $\|f_*^{-1}|_{\Gamma(E^u)}\| < \mu$, and consequently spec f_* does not intersect the unit circle Δ.

Sufficiency. Let spec $f_* = S_1 \cup S_2$, where the sets S_1 and S_2 lie inside and outside the unit circle, respectively. By the theorem on spectral sets ([8], Chapter VII, §3), there exists a decomposition $\Gamma(T_\Lambda M) = \Gamma_1 \oplus \Gamma_2$, where the subspaces Γ_1 and Γ_2 are invariant with respect to f_* and $\operatorname{spec}(f_*|_{\Gamma_i}) = S_i$, $i = 1, 2$.

The space $\Gamma(T_\Lambda M)$ forms a module over the ring $C(\Lambda)$ of continuous functions on Λ. We shall prove that the subspaces Γ_1 and Γ_2 are submodules.

LEMMA 1.1. *If $u(x) \in \Gamma_i$ and $\rho(x) \in C(\Lambda)$, then $\rho(x)u(x) \in \Gamma_i$, $i = 1, 2$.*

PROOF. By the spectral radius theorem, $u(x) \in \Gamma_1$ if and only if

$$\lim_{n\to\infty} \|f_*^n u(x)\|^{1/n} < \lambda < 1.$$

But

$$\|f_*^n\, \rho(x)u(x)\| \leqslant \|\rho(f^{-n}(x))\| \cdot \|f_*^n u(x)\|.$$

Hence

$$\|f_*^n\, \rho(x)u(x)\|^{1/n} \leqslant \|\rho(f^{-n}(x))\|^{1/n} \cdot \|f_*^n u(x)\|^{1/n} < \lambda < 1;$$

that is, $\rho(x)u(x) \in \Gamma_1$. Similarly, the second assertion follows from the fact that $u(x) \in \Gamma_2$ if and only if

$$\lim_{n\to\infty} \|f_*^{-n} u(x)\|^{1/n} < \lambda < 1.$$

LEMMA 1.2. *Let*

$$\Gamma_i(x) = \{\gamma \in T_x M \mid \exists\, u \in \Gamma_i,\ u(x) = \gamma\}, \qquad i = 1, 2.$$

Then $T_*M = \Gamma_1(x) \oplus \Gamma_2(x)$, $\dim \Gamma_i(x)$ *is a continuous (and hence locally constant) function, and the subspaces* $\Gamma_i(x)$ *depend continuously upon the point* x.

PROOF. 1. Let $\gamma \in T_xM$, $u \in \Gamma(T_\Lambda M)$, $u(x) = \gamma$ and $u = u_1 + u_2$, where $u_i \in \Gamma_i$, $i = 1, 2$. Then $\gamma = u_1(x) + u_2(x)$.

2. We now prove that $\Gamma_1(x) \cap \Gamma_2(x) = \varnothing$. Suppose that this is not the case. Let $\gamma_1, \ldots, \gamma_k$ be a basis for $\Gamma_1(x)$, let $\eta_1, \ldots, \eta_l$ be a basis for $\Gamma_2(x)$, let $\gamma_i = l_i(x)$, $l_i \in \Gamma_1$, $i = 1, \ldots, k$, and let $\eta_j = f_j(x)$, $f_j \in \Gamma_2$, $j = 1, \ldots, l$. Let $y \in \Lambda$. Let $\Gamma'_1(y)$ denote the subspace of $T_y\Lambda$ generated by the vectors $l_1(y), \ldots, l_k(y)$, and $\Gamma'_2(y)$ the subspace generated by $f_1(y), \ldots, f_l(y)$. In some neighborhood of x on Λ, these subspaces depend continuously upon y and, consequently, in a neighborhood of x (possibly a smaller one), one can find continuous functions

$$\lambda_1(y), \ldots, \lambda_k(y), \mu_1(y), \ldots, \mu_l(y),$$

$$\sum_{i=1}^{k} |\lambda_i(y)| > 0, \quad \sum_{j=1}^{l} |\mu_j(y)| > 0,$$

such that

$$\sum_{i=1}^{k} \lambda_i(y) l_i(y) = \sum_{j=1}^{l} \mu_j(y) l_j(y).$$

Let $\rho(y)$ be a continuous function on Λ, equal to zero outside this neighborhood and positive at x. Clearly, for any point $y \in \Lambda$,

$$\sum \rho(y)\lambda_i(y)\, l_i(y) = \sum \rho(y)\mu_j(y) f_j(y) = \gamma(y).$$

By virtue of Lemma 1.1, $\gamma(y) \in \Gamma_1 \cap \Gamma_2$. Thus, $\Gamma_1(x) \cap \Gamma_2(x) = \varnothing$.

3. Clearly, $\Gamma'_i(y) \subset \Gamma_i(y)$, $i = 1, 2$. But in some neighborhood of x we have $\dim \Gamma'_1(y) + \dim \Gamma'_2(y) = \dim M$, and therefore, in this neighborhood, $\Gamma'_i(y) = \Gamma_i(y)$, $i = 1, 2$. Thus, the lemma is proved.

LEMMA 1.3.

$$\Gamma_i = \Gamma'_i = \{u \in T_\Lambda M \mid \forall\, x \in \Lambda,\ u(x) \in \Gamma_i(x)\}, \qquad i = 1, 2.$$

PROOF. Clearly, $\Gamma_i \subset \Gamma'_i$. We continue to use the notation of Lemma 1.2. Let $x \in \Lambda$, and let U be a neighborhood of x in Λ such that $\Gamma'_i(y) = \Gamma_i(y)$ for all $y \in U$. Let $u \in \Gamma'_1$, with $u(y) = 0$ for $y \notin U$. Then $u(x) = \Sigma\lambda_i(x) l_i(x)$, where the $\lambda_i(x)$ are continuous functions, and $u \in \Gamma_1$ by Lemma 1.1. We construct such a neighborhood U for each point $x \in \Lambda$. From the covering of Λ obtained in this manner we select a finite subcovering $\{U_1, \ldots, U_m\}$ and consider a partition of unity $\rho_1, \ldots, \rho_m$ associated with this finite covering. If $u \in \Gamma'_1$, then $u = \Sigma_1^m \rho_i u$, $\rho_i u \in \Gamma_1$, and therefore $u \in \Gamma_1$. Similarly, one proves that Γ'_2 and Γ_2 coincide.

We now complete the proof of Theorem 1.1. Letting $E_x^s = \Gamma_1(x)$ and $E_x^u = \Gamma_2(x)$, we show that these subspaces satisfy conditions 1.1.1–1.1.3 of Definition 1.1. Condition 1.1.1 follows from Lemma 1.2 and condition 1.1.2 from Lemma 1.3. We shall verify condition 1.1.3 for $\gamma \in E_x^s$. Let $u \in \Gamma_1$, with $u(x) = \gamma$. By the spectral radius theorem,

$$\lim_{n \to \infty} (\|f_x^n|_{E_x^s}\|)^{1/n} < \mu < 1.$$

Hence

$$\|f_*^n u\| < c\mu^n,$$

$$\|Df^n\gamma\| = \|f_*^n u(f_x^n)\| \leqslant \sup_{y \in \Lambda} \|f_*^n u(y)\| = \|f_*^n u\| < c\mu^n.$$

EXERCISE 1.6. Let Λ be a closed invariant set of f, and suppose that the nonperiodic points of f which lie in Λ are dense in Λ and $1 \notin \operatorname{spec} f_*$. Then Λ is a hyperbolic set of f.

§2. ϵ-trajectories

Fixing some Riemannian metric on the manifold M, we let $\rho(x, y)$ denote the distance between the points x and y induced by this metric.

By an *ϵ-trajectory* of the diffeomorphism $f\colon U \longrightarrow M$, we mean a sequence of points $x_n \in U$ such that $\rho(x_{n+1}, fx_n) < \epsilon$. Clearly, a trajectory of any diffeomorphism sufficiently close to f constitutes an ϵ-trajectory of f. Another example: for the approximate calculation (e.g., by a computer) of the trajectory of the point x_0 under the action of the diffeomorphism f, one actually constructs a sequence $x_0, x_1, \dots, x_N$, merely guaranteeing that $\rho(x_{n+1}, fx_n) < \epsilon$, $\epsilon \approx 10^{-6}$ to 10^{-10}, $n = 0, \dots, N-1$. Since, generally speaking, the error increases exponentially with time, the actual trajectory of the point x_0 will, of course, strongly deviate from this sequence for sufficiently large N (of order 10^3 to 10^4). These examples demonstrate the importance of being able to answer the following question. For a given ϵ-trajectory $\{x_n\}$, can one find a point x such that $\rho(f^n x, x_n)$ is sufficiently small for all n? In other words, can a given ϵ-trajectory be approximated, uniformly in n, by an actual trajectory?

THEOREM ON ϵ-TRAJECTORIES (D. V. ANOSOV). *Let Λ be a hyperbolic set of the diffeomorphism f. Then there exists a neighborhodd $U(\Lambda) \supset \Lambda$ such that for any $\delta > 0$ there is an $\epsilon > 0$ such that to any ϵ-trajectory $\{x_n\} \subset U(\Lambda)$ there corresponds a trajectory $\{f^n x\}$ with $\rho(x_n, f^n x) < \delta$, $n \in \mathbf{Z}$.*

This theorem follows from a more general assertion, also due to Anosov [9], which also subsumes all known propositions concerning the local stability of hyperbolic sets. Roughly speaking, this assertion is as follows. If in a small neighborhood of a hyperbolic set of the diffeomorphism f there is given not one but a family of ϵ-trajectories, and also, on the point set formed by these ϵ-trajectories, a topology at least as strong as the topology induced by M, then one can, while preserving the given topology, approximate this family by a family of actual trajectories of any diffeomorphism sufficiently close to f.

THEOREM ON FAMILIES OF ϵ-TRAJECTORIES. *Let Λ be a hyperbolic set of the diffeomorphism $f\colon U \longrightarrow M$. Then there exist a neighborhood $U(\Lambda)$ of Λ and numbers $\epsilon_0, \delta_0 > 0$ with the following properties. For any $\delta > 0$ there is an $\epsilon > 0$ such that, for any diffeomorphism $f'\colon U(\Lambda) \longrightarrow M$ with $\rho_{C^1}(f, f') < \epsilon_0$, any topological space X, any homeomorphism $g\colon X \longrightarrow X$ and any continuous map $\varphi\colon X \longrightarrow U(\Lambda)$ for which*

$$P(\varphi g, f'\varphi) = \sup_{x\in X} \rho(\varphi g(x),\ f'\varphi(x)) < \epsilon,$$

one can find a continuous map $\psi\colon X \longrightarrow U(\Lambda)$ such that (1) $\psi g = f'\psi$, *and* (2) $P(\varphi, \psi) < \delta$.

Moreover, if for some map ψ' we have $\psi' g = f'\psi'$ and $P(\varphi, \psi') < \delta_0$, then $\psi' = \psi$.

PROOF. Let $C(X, M)$ denote the metric space of all continuous maps of X into M, with the metric P, and let $C_\varphi(X, TM)$ be the Banach space of continuous mappings $u\colon X \longrightarrow TM$ such that $u(x) \in T_{\varphi(x)}$, $x \in X$, with the norm $\|u\| = \sup_{x\in X}\|u(x)\|$. There exists a number $\theta > 0$, depending only upon the choice of the Riemannian metric on M, and independent of X, g and φ, such that the mapping Φ, given by

$$(\Phi\psi)(x) = \exp^{-1}_{\varphi(x)}\psi(x),\quad x \in X,\ \psi \in C(X, M)$$

maps the ball in $C(X, M)$ with center φ and radius θ homeomorphically onto the ball

$$V_\varphi = \{v \in C_\varphi(X, TM) \mid \|v\| < \theta\}.\text{(2)}$$

We define a mapping $F\colon\ C(X, U) \longrightarrow C(X, M)$ by the formula $F\varphi = f'\varphi g^{-1}$. Fix $\varphi \in C(X, U)$, and let

$$F^\varphi = \Phi F \Phi^{-1}\colon V_\varphi \longrightarrow C_\varphi(X, TM),\qquad (F^\varphi v)(x) = \exp^{-1}_{\varphi(x)} f' \exp_{\varphi(g^{-1}x)} v(g^{-1}x),$$

$v \in V_\varphi$, $x \in X$. This mapping is differentiable. Direct calculation shows that for $v \in V_\varphi$ and $u \in C_\varphi(X, TM)$

$$[(DF^\varphi)_V u]\,(x) \tag{2.1}$$

$$= (D\exp^{-1}_{\varphi(x)})_{f'\exp_{\varphi(g^{-1}x)}v(g^{-1}x)}\ [Df'(D\exp_{\varphi(g^{-1}x)})_{V(g^{-1}x)}u(g^{-1}x)].$$

From this formula it is clear that the operator $(DF^\varphi)_V$ satisfies a Lipschitz condition with respect to V, with a Lipschitz constant K which depends continuously on f' in

(2) The mappings Φ constructed for all the various φ are consistent and define a Banach manifold structure on $C(X, M)$ (see [10], Appendix III to the Russian transl., §2).

the C^1-topology, and also depends on the exponential map in M (i.e. on the Riemann metric), but does not depend on X, g or φ. Setting $v = 0$ in (2.1), we obtain

$$((Df^{\varphi})_0 u)(x) = (D\exp^{-1}\varphi(x))_{f'\varphi(g^{-1}x)} Df'u(g^{-1}x). \tag{2.2}$$

There is a neighborhood $U_1(\Lambda) \supset \Lambda$ such that the decomposition $T_xM = E_x^s \oplus E_x^u$, defined for $x \in \Lambda$, can be continuously extended to all $x \in U_1(\Lambda)$. Suppose that the Riemannian metric on M is Ljapunov with respect to E. Let P_x^s denote the operator projecting T_xM onto E_x^s along the direction parallel to E_x^u, and let $P_x^u = \mathrm{Id} - P_x^s$. Choose a neighborhood $U(\Lambda)$, $\Lambda \subset U(\Lambda) \subset U_1(\Lambda)$, a neighborhood $\mathfrak{N}$ of the diffeomorphism f in the space of C^1-mappings of $U(\Lambda)$ into M, and a number $\mu < 1$ such that, for $f' \in \mathfrak{N}$, $x \in U(\Lambda)$, and $u \in E^s_{f'(x)}$

$$\|P^s_{f'(x)} Dfu\| < \mu \|u\|,$$

$$\|P^u_{f'(x)} Dfu\| < \frac{\mu}{100}\|u\|,$$

and for $u \in E^u_{f'(x)}$

$$\|P^u_{f'(x)} Dfu\| > \mu^{-1}\|u\|,$$

$$\|P^s_{f'(x)} Dfu\| < \frac{\mu}{100}\|u\|.$$

Since, for any point $y \in M$, $(D\exp_y^{-1})_0$ is the identity operator, it follows that for any $\theta > 0$ one can find an $\eta(\theta)$ such that $\rho(y_1, y_2) < \eta(\theta)$ $(y_1, y_2 \in M)$ implies

1) $1 - \theta < \|(D\exp_{y_1}^{-1})_{y_2} v\|/\|v\| < 1 + \theta, v \in T_{y_2}M$,

2) $\|P^u_{y_1}(D\exp_{y_1}^{-1})_{y_2} v\| < \theta,\ v \in E^s_{y_2}$, and

3) $\|P^s_{y_1}(D\exp_{y_1}^{-1})_{y_2} v\| < \theta,\ v \in E^u_{y_2}$.

Let $\varphi \in C(M, U(\Lambda))$, and set

$$B^s = \{u \in C_\varphi(X, TM): \forall x \in X\ u(x) \in E^s_{\varphi(x)}\},$$
$$B^u = \{u \in C_\varphi(X, TM): \forall x \in X\ u(x) \in E^u_{\varphi(x)}\}.$$

Clearly, B^s and B^u are closed subspaces of $C_\varphi(X, TM)$, and

$$C_\varphi(X, TM) = B^u \oplus B^s. \tag{2.3}$$

This decomposition permits the representation of the operator DF_0^φ in the form

$$DF_0^\varphi = \begin{pmatrix} A^{ss} & A^{su} \\ A^{us} & A^{uu} \end{pmatrix},$$

where A^{ss}: $B^s \longrightarrow B^s$, etc. By virtue of the choice of the neighborhood $U(\Lambda)$ and properties 1)–3), one can determine a number $\theta > 0$ and a neighborhood $\mathfrak{N}_1$ of f in the space $C'(U(\Lambda), M)$, depending on the Riemannian metric in M, such that for

$f' \in \mathfrak{R}_1$ and $P(\varphi g, f'\varphi) < \eta(\theta)$ the following inequalities hold:

$$\|A^{ss}\| < \frac{1+\mu}{2}, \qquad \|A^{su}\| < \frac{\mu}{10},$$

$$\|A^{uu}\| > \left(\frac{1+\mu}{2}\right)^{-1}, \qquad \|A^{us}\| < \frac{\mu}{10},$$

i.e., for any $\lambda \in \Delta$,

$$\|(DF^{\varphi})_0 - \lambda \mathrm{Id})^{-1}\| < R, \tag{2.4}$$

where R is independent of X, g and φ. The neighborhood $\mathfrak{R}_1$ allows us to define the number ϵ_0.

We require this estimate not only to complete the proof of the theorem, but also in the ensuing proof of the roughness of the hyperbolicity property.

We now proceed to the final stage of the proof of the present theorem.

If $u \in V_{\varphi}$ is a fixed point of the mapping F^{φ}, then $\psi = \Phi^{-1}u$ is a fixed point of F, and therefore $\psi = f'\psi g^{-1}$; that is, $\psi g = f'\psi$.

Let $F^{\varphi}u = (DF^{\varphi})_0 u + Hu$. To determine a fixed point, we must solve the equation

$$((DF^{\varphi})_0 - \mathrm{Id})u = -Hu$$

or $u = Tu$, where

$$Tu = -((DF^{\varphi})_0 - \mathrm{Id})^{-1} Hu.$$

Since DF^{φ} satisfies a Lipschitz condition with constant K, we have

$$\|T_{u_1} - T_{u_2}\| < RK \max(\|u_1\|, \|u_2\|)\|u_1 - u_2\|.$$

Furthermore,

$$\|T(0)\| < R\|H(0)\| < RNP(\varphi, f'\varphi g^{-1}),$$

where the constant N depends only upon the Riemannian metric on M.

We choose a number θ_1 such that $0 < \theta_1 < \min(\theta, \delta/2, (4RK)^{-1})$. Then, for $u_1, u_2 \in C_{\varphi}(X, TM)$, $\|u_1\| < \theta_1$ and $\|u_2\| < \theta_1$, we have

$$\|Tu_1 - Tu_2\| < \tfrac{1}{2}\|u_1 - u_2\|.$$

Now let $\epsilon = \theta_1/2RN$ and let $\varphi \in C(X, M)$ be such that $\operatorname{Im} \varphi \subset U(\Lambda)$ and

$$P(\varphi, f'\varphi g^{-1}) = P(\varphi g, f'\varphi) < \epsilon.$$

Then $\|T(0)\| < \theta/2$, and a well-known fixed-point theorem (see [11], Theorem 10.12) can be applied to the ball $V_{\theta_1} = \{u \in C_{\varphi}(X, TM) | \, \|u\| < \theta_1\}$ and the mapping T, whence it follows that there is a unique fixed point u of T in V_{θ_1}. Therefore $\psi = \Phi u$ is the mapping whose existence is asserted in the theorem. Finally, we note that one can assert the uniqueness of the fixed point in some neighborhood of φ, the size of which is independent of δ, since the condition $\theta_1 < 1/4RK$, which guarantees uniqueness, does not depend upon δ.

This completes the proof of the theorem.

THEOREM 2.1 (Roughness of the property of hyperbolicity). *Let Λ be a hyperbolic set of the diffeomorphism f: $U \longrightarrow M$. Then there exist a neighborhood $U(\Lambda)$ of Λ and a number $\epsilon_0 > 0$ such that any closed invariant set $\Gamma \subset U(\Lambda)$ of any diffeomorphism f': $U(\Lambda) \longrightarrow M$, $\rho_{C'}(f, f') < \epsilon_0$, is hyperbolic.*

PROOF. Consider Γ as a topological space with the homeomorphism $f'|_\Gamma$ and the imbedding φ: $\Gamma \longrightarrow M$. The operator f_* defined in §1.4 coincides, in this case, with $(DF^\varphi)_0$. The estimate (2.4) in the proof of the theorem on families of ϵ-trajectories shows that, for an appropriate choice of $U(\Lambda)$ and ϵ_0, $(\text{spec } f_*) \cap \Delta = \varnothing$, and therefore, by Theorem 1.1, Γ is a hyperbolic set.

COROLLARY 2.1. *The Anosov diffeomorphisms form an open set in* $\text{Diff}^1(M)$.

COROLLARY 2.2. *Under the conditions of the theorem on families of ϵ-trajectories, the closure of the invariant set $\psi(X)$ is a hyperbolic set for f'.*

2. Consequences of the theorem on families of ϵ-trajectories.

1. THEOREM 2.2. *Let $x \in \Omega(f|_\Lambda)$. Then, in any neighborhood V of x, there is a periodic point of the diffeomorphism f.*

PROOF. One may assume that V is a ball of radius δ and that $V \subset U(\Lambda)$. In the theorem on families of ϵ-trajectories, we choose an ϵ corresponding to this δ, and denote by $V_{\epsilon/2}$ the intersection of Λ with the $\epsilon/2$-neighborhood of x. We take an $N > 0$ such that $f^{-N}V_{\epsilon/2} \cap V_{\epsilon/2} \neq \varnothing$, and let x' be a point of this intersection. Consider the space $X_N = \{0, 1, \ldots, N-1\}$, with the discrete topology, the homeomorphism $g(k) = k + 1 \pmod N$, and the mapping φ: $X_N \longrightarrow M$ defined by $\varphi k = f^k x'$, $k = 0, \ldots, N-1$.

Clearly this situation satisfies the conditions of the theorem on ϵ-trajectories, since

$$P(\varphi g, f\varphi) = \rho(f^N x', x') < \epsilon.$$

Consequently there is a mapping ψ such that $\psi(0) \in V$ and $\psi(0)$ is a periodic point of f. In general, the mapping ψ need not be injective, and therefore the period of this trajectory may be less than N.

COROLLARY 2.3. *The periodic points of an Anosov diffeomorphism f: $M \longrightarrow M$ are dense in the set $\Omega(f)$ of nonwandering points.*

2. THEOREM 2.3 (Stability of a hyperbolic set). *Let Λ be a hyperbolic set of the diffeomorphism f: $U \longrightarrow M$. For any neighborhood V, $\Lambda \subset V \subset U$, and any*

$\delta > 0$, there is an $\epsilon > 0$ such that any diffeomorphism f': $V \longrightarrow M$ for which $\rho_{C^1}(f', f|_V) < \epsilon$ has an invariant hyperbolic set $\Lambda(f') \subset V$ which admits a homeomorphism h: $\Lambda(f') \longrightarrow \Lambda$ such that $h \cdot f'|_{\Lambda(f')} = f|_\Lambda \cdot h$.

PROOF. We apply the theorem on families of ϵ-trajectories to the topological space Λ (considered as a space in itself, rather than a subset of M), the homeomorphism $f|_\Lambda$: $\Lambda \longrightarrow \Lambda$ and the imbedding φ: $\Lambda \longrightarrow M$. This permits us to construct a mapping ψ: $\Lambda \longrightarrow M$ such that $\psi \cdot f|_\Lambda = f' \cdot \psi$. We denote the image of ψ by $\Lambda(f')$. If ψ is a homeomorphism of Λ onto $\Lambda(f')$, it is then sufficient to let $h = \varphi \cdot \psi^{-1}$, and the theorem is proved. Now, applying the theorem on families of ϵ-trajectories to the topological space $\Lambda(f')$, the homeomorphism $f'|_{\Lambda(f')}$ and the imbedding φ': $\Lambda(f') \longrightarrow M$, we construct a mapping ψ': $\Lambda(f') \longrightarrow M$ such that $\psi' \cdot f'|_{\Lambda(f')} = f \cdot \psi'$. Clearly, $\psi' \cdot (\varphi')^{-1} \cdot \psi \cdot f|_\Lambda = f \cdot \psi' \cdot (\varphi')^{-1} \cdot \psi$ and $P(\varphi, \psi' \cdot (\varphi')^{-1} \cdot \psi)$ is sufficiently small, provided ϵ is chosen sufficiently small. Therefore, by the uniqueness assertion of the theorem on families of ϵ-trajectories, $\psi' \cdot (\varphi')^{-1} \cdot \psi = \varphi$ and $\varphi^{-1} \cdot \psi' \cdot (\varphi')^{-1} \cdot \psi$ is the identity mapping of Λ. It follows that ψ is a homeomorphism. The hyperbolicity of $\Lambda(f')$ follows from Theorem 2.1.

COROLLARY 2.4. *An Anosov diffeomorphism is structurally stable.*

3. The theorem on ϵ-trajectories corresponds to the assertion regarding the existence of the mapping ψ in the theorem on families of ϵ-trajectories, applied to the following situation:

$$X = \mathbf{Z}, \quad g(n) = n + 1, \quad n \in \mathbf{Z}, \quad \varphi(n) = x_n.$$

The assertion regarding uniqueness in this situation enables us to obtain the following useful fact.

THEOREM 2.4 ("Expansiveness" of a hyperbolic set). *There is an $\epsilon > 0$ such that $y_1, y_2 \in \Lambda$ $(y_1 \neq y_2)$ implies that, for some integer n,*

$$\rho(f^n y_1, f^n y_2) \geqslant \epsilon. \tag{2.5}$$

REMARK. One may assume that $y_1 \in \Lambda$ and $y_2 \in U(\Lambda)$, and the assertion of the theorem will still remain valid.

§3. Stable and unstable manifolds

1. A considerable portion of the results relating to hyperbolic sets is based upon the possibility of "projecting" the subspaces E_x^s and E_x^u (or, at least, small balls in these subspaces) onto the manifold M in a manner invariant with respect to the diffeomorphism. In the case of hyperbolic fixed (or periodic) points, this operation is described by the Hadamard-Perron theorem, which is formulated and roughly proved in §5 of Kušnirenko's lectures. In those lectures, two more situations of interest to us were considered. At the end of §5 (the proof of the Grobman-Hartman theorem), for any diffeomorphism φ: $\mathbf{R}^2 \longrightarrow \mathbf{R}^2$ which is uniformly C^1-close to a hyperbolic

linear transformation, there are constructed families of smooth curves, invariant with respect to φ, and close to families of straight lines which run parallel to the eigenvectors of the linear transformation. In doing this, one first constructs expanding and contracting direction fields which are invariant with respect to $D\varphi$; that is, one essentially proves that $\mathbf{R}^2$ is a hyperbolic set of the diffeomorphism φ (except that it is noncompact). In the second instance, similar considerations lead to the theorem on constantly acting perturbations (§8.3), which is formulated for a sequence of diffeomorphisms of $\mathbf{R}^n$ into itself which are C^1-close to one and the same hyperbolic linear transformation. Incidentally, we remark that the previous assertion concerning the existence of families of curves follows from this theorem, with the exception of the part regarding the smoothness (higher than C^1) of the curves. It turns out that the theorem on constantly acting perturbations is valid in the more general situation where there is given a sequence of diffeomorphisms C^1-close to a sequence of linear transformations A_m: $\mathbf{R}^n \longrightarrow \mathbf{R}^n$, $m \in \mathbf{Z}$, hyperbolic in the following sense: there exist a decomposition $\mathbf{R}^n = \Gamma^s_m \oplus \Gamma^u_m$ and a number $\mu < 1$ such that

$$A_m\Gamma^s_m = \Gamma^s_{m+1}, \quad A\Gamma^u_m = \Gamma^u_{m+1}, \quad \|A_m u\| < \mu\|u\|, \quad u \in \Gamma^s_m,$$
$$\|A_m v\| > \mu^{-1}\|v\|, \quad v \in \Gamma^u_m.$$

From this theorem on constantly acting perturbations, it is not difficult to derive the following theorem, which is very important in the study of hyperbolic sets.

THEOREM ON STABLE MANIFOLDS. *Let Λ be a hyperbolic set of the C^1-diffeomorphism f: $U \longrightarrow M$, and ρ a metric on M induced by a Ljapunov-Riemannian metric. Then there is an $\epsilon_0 > 0$ such that, for $0 < \epsilon < \epsilon_0$ and $x \in \Lambda$, there exists a mapping*

$$\varphi_{x,\epsilon}\colon \mathbf{D}^{k(x)} \longrightarrow M \quad (k(x) = \dim E^s_x)$$

with the following properties:

P.1. *$\varphi_{x,\epsilon}$ is a C^1-imbedding of $\mathbf{D}^k$ in M (we denote the image of $\varphi_{x,\epsilon}$ by $W^s_\epsilon(X)$).*

P.2. $T_x W^s_x(x) = E^s_x$.

P.3. *$f(W^s_\epsilon(x)) \subset W^s_\epsilon(f(x))$, and $\rho(f^n x, f^n y) \longrightarrow 0$ as $n \longrightarrow \infty$ for $y \in W^s_\epsilon(x)$.*

P.4. *The correspondence $x \longrightarrow \varphi_{x,\epsilon}$ defines a continuous mapping $\Lambda \longrightarrow$ $C^1(\mathbf{D}^k, M)$.*

P.5. $W^s_\epsilon(x) = \{y \in M \mid \rho(f^n y, f^n x) < \epsilon, n \geqslant 0\}$.

Substituting f^{-1} for f in this theorem, we obtain an analogous assertion concerning the existence of local unstable manifolds $W^u_\epsilon(x)$. In what follows we shall generally speak only of contracting manifolds, since all the corresponding definitions and assertions concerning expanding manifolds are easily obtained by substituting f^{-1} for f.

2. We shall show how this theorem is reduced to the theorem on constantly acting perturbations. We fix an $\epsilon_1 > 0$ such that, for any point $x \in M$, the mapping $\exp_x^{-1}$ maps the ball with radius ϵ_1 and center x in M diffeomorphically onto the ball $D_\epsilon(x) = \{u \in T_x M \mid \|u\| < \epsilon\}$. Let $\epsilon_2 = \epsilon_1(\max_{x \in M}\|Df_x\|)^{-1}$. Fix a point $x \in M$,

and consider the sequence of mappings

$$f_n : D_{\epsilon_2}(T_{f^{n-1}(x)}M) \longrightarrow D_{\epsilon_1}(T_{f^n(x)}M),$$

where

$$f_n u = \exp^{-1}_{f^n(x)} f \exp_{f^{n-1}(x)} u.$$

Clearly, $f_n(0) = 0$ and $(Df_n)_0 = Df_{f^{n-1}(x)}$.

If we restrict the mapping f_n to sufficiently small balls, and then extend these restrictions to mappings $\widetilde{f}_n$: $T_{f^{n-1}(x)}M \longrightarrow T_{f^n(x)}M$ such that the $\widetilde{f}_n$ are uniformly (with respect to n) C^1-close to the $Df_{f^{n-1}(x)}$, we encounter a situation conforming with the conditions of the theorem on constantly acting perturbations, and statements P.1–P.5 follow from the assertion of this theorem.

We shall not prove the theorem on constantly acting perturbations. We note that in Kušnirenko's lectures a rough sketch of one of the possible methods of proof is given, following the statement of the theorem. A property corresponding to P.5 constitutes the basis for this method of proof.

Another method of proof is based upon a successive approximation process which is used in Kušnirenko's lectures for the proof of the Hadamard-Perron and Grobman-Hartman theorems, and is described there with all essential details. For this method of proof, properties P.1–P.4 are verified in a simple manner. Moreover, these properties remain valid in a more general situation (see §7) where property P.5 may no longer hold. We note that detailed proofs of the theorem on stable manifolds for hyperbolic sets are contained in Chapter 1, §6, of the dissertation of V. M. Alekseev [12] (his method is based upon property P.5), and in the article of Hirsch and Pugh [13] (another method, apparently not the most straightforward, but quite elegant, based upon the reduction of the theorem on stable manifolds to the Hadamard-Perron theorem in a Banach space). The proof given by Hirsch and Pugh is also presented in Nitecki's book [14].

3. One may define, not only the local stable manifold $W^s_\epsilon(x)$, but also the global stable manifold (in general, not connected)

$$W^s(x) = \bigcup_{n=0}^{\infty} f^{-n} W^s_\epsilon(f^n x) = \{y \in M | \rho(f^n y, f^n x) \longrightarrow 0 \text{ as } n \longrightarrow +\infty\}.$$

If f acts on the entire manifold M, then there exists an immersion φ: $\mathbf{R}^k \longrightarrow M$ such that Im $\varphi = W^s(x)$. This immersion establishes a one-to-one correspondence between $\mathbf{R}^k$ and $W^s(x)$, but the mapping φ^{-1} is, in general, discontinuous (that is, φ is not an imbedding). In other words, the topology induced on $W^s(x)$ as a subset of M is weaker (i.e., has fewer open sets) than the topology induced by φ. We remark that $W^s_\epsilon(x)$ may be described as the arcwise connected component, containing x, of the intersection of $W^s(x)$ and the ϵ-neighborhood of x.

4. The stable and unstable manifolds of the Anosov diffeomorphism $f_A\colon \mathbf{T}^2 \longrightarrow \mathbf{T}^2$ in Example 1.2 are described in Kušnirenko's lectures (§8, Example 12). Let $x \in \mathbf{T}^2$. Then $W^s(x)$ is the projection onto $\mathbf{T}^2$ of the line in $\mathbf{R}^2$ parallel with the eigenvector e_1 and passing through any point y such that $\pi y = x$. Alternatively, $W^s(x) = \exp_x(E_x^s)$, provided the Riemannian metric on $\mathbf{T}^2$ is invariant with respect to group translations. The manifold $W^s(x)$ is dense in $\mathbf{T}^2$. Thus the immersion φ described in the previous subsection is, in fact, not an imbedding.

A similar situation arises in Examples 1.3 and 1.4. Here, also, $W^s(x) = \exp_x E_x^s$.

EXERCISE 3.1. Prove that every manifold $W^s(x)$ is dense in $\mathbf{T}^n$ (or N/Γ, as the case may be).

EXERCISE 3.2. Let f and f' be Anosov diffeomorphisms of the manifold M, and $h\colon M \longrightarrow M$ a homeomorphism such that $f \cdot h = h \cdot f'$. Prove that $h(W_{f'}^s(x)) = W_f^s(hx)$ (here the subscripts f and f' indicate the diffeomorphism with respect to which the stable manifold is formed).

We now turn to Example 1.5. Clearly, in this case the local stable manifold $W_\epsilon^s(x)$ of the point $x \in \Lambda$ is the intersection of the ϵ-neighborhood of x with the vertical segment $\Delta^{i_{-1}\cdots i_{-n}}$ containing this point. The global stable manifold $W^s(x)$ is not connected.

EXERCISE 3.3. $W^s(x) = \bigcup_{s=0}^{\infty} \bigcup_{j_1,\dots,j_s} \Delta^{j_1\cdots j_s i_{-s-1} i_{-s-2}\cdots}$.

EXERCISE 3.4. Extend the diffeomorphism f to a diffeomorphism $S^2 \longrightarrow S^2$. Then the extended manifold $W^s(x)$ is the image of a line under an immersion, and therefore the segments $\Delta^{j_1\cdots j_s i_{-s-1} i_{-s-2}\cdots}$ are ordered according to the order of their inverse images on the line. Describe this ordering.

One of the methods of extending f to a diffeomorphism $S^2 \longrightarrow S^2$ is illustrated in Figure 4. Enlarge the square to a smooth disk by adding two ends, and extend f to a mapping of the disk into itself such that the point P is attractive. Then extend this mapping to the complement of this disk in S^2 in such a way that there is a repulsive fixed point Q outside the disk. In this case, $\Omega(f) = \{P\} \cup \{Q\} \cup \Lambda$ (see [2], §1.5).

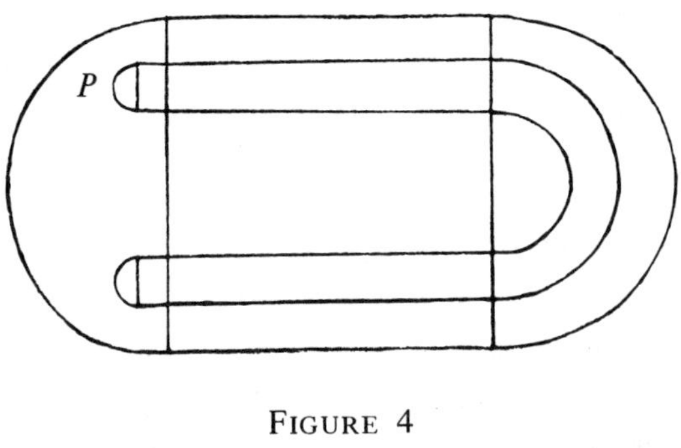

FIGURE 4

5. The theorem on stable manifolds will be used intensively in the succeeding sections. Along with the theorem on families of ϵ-trajectories and its corollaries, this is the

most important technical statement in the theory of hyperbolic sets. We shall now formulate several lemmas on stable and unstable manifolds, which will play a role in §§4–6. These lemmas follow in a straightforward manner from the theorem on stable manifolds. The reader may regard the proofs as useful exercises. We consider M, f, Λ and the Ljapunov Riemannian metric on M as fixed.

LEMMA 3.1. *There exists an $\epsilon_1 > 0$ such that, for any two points $x_1, x_2 \in \Lambda$, the intersection $W^s_{\epsilon_1}(x_1) \cap W^u_{\epsilon_1}(x_2)$ consists of no more than a single point and is transversal.*

This intersection, if nonempty, will be denoted by $r(x_1, x_2)$.

LEMMA 3.2. *For any $\delta > 0$, there is an $\epsilon_2(\delta) > 0$ such that if $x_1, x_2 \in \Lambda$ and $\rho(x_1, x_2) < \epsilon_2(\delta)$, then*

$$W^s_\sigma(x_1) \cap W^u_\sigma(x_2) \neq \varnothing.$$

We denote by ρ^s (ρ^u) the intrinsic metric on any manifold $W^s(x)$ ($W^u(x)$). Obviously, $\rho^s(x_1, x_2) \geqslant \rho(x_1, x_2)$ for any $x_1, x_2 \in W^s(x)$.

LEMMA 3.3. *For any $\delta > 0$, there is an $\epsilon_3(\delta)$ such that if $x \in \Lambda$, $x_1, x_2 \in W^s_{\epsilon_1}(x)$ and $\rho(x_1, x_2) < \epsilon_3(\delta)$, then*

$$\rho^s(x_1, x_2) < (1 + \delta)\rho(x_1, x_2).$$

LEMMA 3.4. *There exist numbers $\epsilon_4 > 0$ and $\mu < 1$ such that if $x \in \Lambda$ and $x_1, x_2 \in W^s_{\epsilon_4}(x)$, then*

$$\rho^s(f(x_1), f(x_2)) < \mu\rho^s(x_1, x_2).$$

COROLLARY 3.1. *If $f\colon M \longrightarrow M$ is an Anosov diffeomorphism, $x \in M$ and $x_1, x_2 \in W^s(x)$, then*

$$\rho^s(f(x_1), f(x_2)) < \mu\rho^s(x_1, x_2),$$

where $\mu < 1$ is independent of x, x_1 and x_2.

LEMMA 3.5. *There exist numbers $K > 0$ and $\epsilon_5 > 0$ such that if $\rho(x_1, x_2) < \epsilon_5$, then $W^s_{\epsilon_1}(x_1) \cap W^u_{\epsilon_1}(x_2) \neq \varnothing$ and*

$$\rho^s(x_1, r(x_1, x_2)) + \rho^u(r(x_1, x_2), x_2) < K\rho(x_1, x_2).$$

EXERCISE 3.5. Using the theorem on stable manifolds and Lemmas 3.1–3.5, but not the theorem on families of ϵ-trajectories, prove Theorem 2.4 along with the following corollary: *If $y_2 \notin W^s_\epsilon(y_1)$, then (2.5) holds for some $n \geqslant 0$.*

6. In the study of hyperbolic sets, the techniques of ϵ-trajectories and stable manifolds complement one another and are, in part, interchangeable. In §2, we singled out that part of the theory of hyperbolic sets which can be constructed without using the theorem on stable manifolds. At the same time, many of the results appearing in

§2 are ordinarily obtained by the use of stable and unstable manifolds (the first proof of the structural stability of Anosov diffeomorphisms ([1], §10), Theorem 2.4 on the "expansiveness" of hyperbolic sets, the density of periodic points, etc.). We shall now show that even a substantial part of the theorem on families of ϵ-trajectories, covering many important applications, can be proved by means of a successive approximation process which makes use of stable and unstable manifolds and does not depend upon the Riemannian metric. Here we are referring to the spectral case of this theorem in which $f' = f$ and $\operatorname{Im} \varphi \subset \Lambda$. We now present this proof, omitting only certain technical details.

Thus, let $\varphi\colon X \longrightarrow \Lambda$ be a mapping for which $\rho(\varphi, F\varphi)$ is sufficiently small, $\varphi_0 = \varphi$, and

$$\varphi_n(x) = F^{-1}(W^s_{\epsilon_1}(\varphi'_{n-1}(x)) \cap W^u_{\epsilon_1})\,((F\varphi'_{n-1})(x)),$$

where

$$\varphi'_{n-1} = W^s_{\epsilon_1}(\varphi_{n-1}(x)) \cap W^u_{\epsilon_1}((F\varphi_{n-1})\,(x))$$

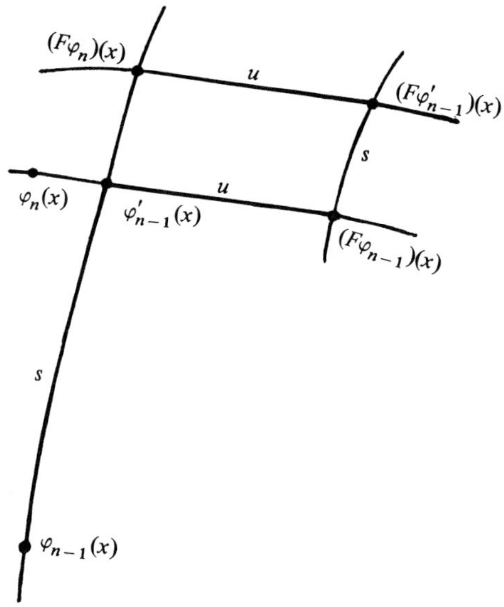

FIGURE 5

(see Figure 5); the number ϵ_1 is defined in Lemma 3.1. For purposes of calculation, it is convenient, as a measure of the mutual deviation of two nearby points $y_1, y_2 \in M$, to use the quantity

$$\begin{aligned}\overline{\rho}(y_1, y_2) = \rho^s(y_1, r(y_1, y_2)) + \rho^u(r(y_1, y_2), y_2)\\ + \rho^s(y_2, r(y_2, y_1)) + \rho^u(r(y_2, y_1), y_1),\end{aligned}$$

which is not a metric, but is related to ρ by the inequality $2\rho < \bar{\rho} < R\rho$, where R depends only upon the choice of the Riemannian metric (see Lemma 3.5), and to estimate the corresponding deviation of two close (in $C(X, M)$) mappings $\psi_1, \psi_2\colon X \longrightarrow M$ by the quantity

$$\bar{P}(\psi_1, \psi_2) = \sup_{x \in X} \bar{\rho}(\psi_1(x), \psi_2(x)).$$

It is not difficult to show that, for sufficiently small $\bar{P}(\varphi, F(\varphi))$,

$$\bar{P}(\varphi_n, F(\varphi_n)) < \mu \bar{P}(\varphi_{n-1}, F(\varphi_{n-1})),$$

where $\mu < 1$, and

$$\bar{P}(\varphi_n, \varphi_{n-1}) < R_1 \bar{P}(\varphi_{n-1}, F(\varphi_{n-1})).$$

From this it follows that $\psi = \lim \varphi_n$ is a fixed point of the mapping ψ, and

$$P(\varphi, \psi) < \frac{RR_1}{i - \mu} \bar{P}(\varphi, F(\varphi)),$$

§4. Anosov diffeomorphisms

In §1 we described most known examples of Anosov diffeomorphisms, and in §3 we proved a series of general assertions concerning Anosov diffeomorphisms, which are valid without any further assumptions. We state these assertions once again:

The Anosov diffeomorphisms form an open set (possibly empty) in the space $\mathrm{Diff}^1(M)$ (Corollary 2.1).

The closure of the set of periodic points of an Anosov diffeomorphism coincides with the set of nonwandering points (Corollary 2.3).

Anosov diffeomorphisms are structurally stable (Corollary 2.4).

In the case of Anosov diffeomorphisms, one may add certain remarks and complements to the theorem on stable manifolds. In fact, in this case the totality of subspaces E_x^s $(x \in M)$ represents a subbundle of the tangent bundle, or a k-dimensional distribution (i.e. a field of k-dimensional tangent planes on M). In this case the theorem on stable manifolds means that this distribution is integrable. The totality of integral manifolds $W^s(x)$ of this distribution constitutes a foliation, which we denote by W^s. As is well known, the integrability conditions of a smooth distribution are given by the theorem of Frobenius. But the distribution E^s may not be smooth (such examples appear in §24 of Anosov's monograph [1]; recall that, in our definition, we did not even assume that E^s is continuous). Therefore, the theorem on stable manifolds cannot be derived from the Frobenius theorem.

The following assertion can be deduced in a straightforward manner from the theorem on stable manifolds.

THEOREM 4.1 (local product structure). *Let f: $M \longrightarrow M$ be an Anosov diffeomorphism, and let $x \in M$. Then there exist a number $\epsilon > 0$, a neighborhood U of x, and a homeomorphism φ: $U \longrightarrow B^n = \{t_1, \ldots, t_n | |t_i| < 1, i = 1, \ldots, n\}$ such that*

$$\varphi(x) = 0, \quad \varphi(y) = (\varphi_1(y), \ldots, \varphi_n(y)),$$

$$\varphi(W^s_{\epsilon_1}(x) \cap U) = \{(\varphi_1(y), \ldots, \varphi_k(y), t_{k+1}, \ldots, t_n), y \in U\},$$

$$\varphi(W^u_{\epsilon_1}(x) \cap U) = \{(t_1, \ldots, t_k, \varphi_{n-k+1}(y), \ldots, \varphi_n(y)), y \in U\}.$$

In other words, the local stable manifolds are transformed into sections of the cube B^n by k-dimensional hyperplanes parallel to the first k coordinate axes, and the local unstable manifolds into sections by l-dimensional hyperplanes parallel to the remaining l coordinate axes.

We remark that if one only requires that the leaves of one foliation be transformed into hyperplane sections, then the homeomorphism φ can be chosen so as to be differentiable along the directions of this foliation.

A more essential result is the following complement to the theorem on stable manifolds.

THEOREM 4.2 (uniqueness of integral manifolds). *Any curve γ of class C^1 on M, which is tangent to the subspace E^s_x at every point $x \in \gamma$, belongs to one leaf of $W^s(x)$.*

PROOF. Let the metric on M be Ljapunov and let the length $l(\gamma)$ of γ be less than $\epsilon/2$, where ϵ is chosen as in Theorem 2.4. If γ is tangent to E^s_x at every point, then, clearly, $l(f^n\gamma) < \mu^n l(\gamma) < \epsilon/2$, $\mu < 1$. But if $x_1, x_2 \in \gamma$ and $x_2 \notin W^s_\epsilon(x_1)$, then there is an $n > 0$ such that $\rho(f^n x_1, f^n x_2) \geqslant \epsilon$ (see Exercise 3.5), and therefore $l(f^n\gamma) > \epsilon$. Thus, $\gamma \subset W^s_\epsilon(x)$ for any point $x \in \gamma$. Now, decomposing an arbitrary curve γ into segments of length less than $\epsilon/2$, we obtain the assertion of the theorem.

2. Further results relating to Anosov diffeomorphisms are based upon the additional assumption

$$\Omega(f) = M. \tag{4.1}$$

THEOREM 4.3. *Let f: $M \longrightarrow M$ be an Anosov diffeomorphism. Then the following assertions are true*:

4.3.1. *The periodic points of f are dense in M.*

4.3.2. *f is topologically transitive.*

4.3.3. *Any leaf of the foliation W^s (or W^u) is dense in M.*

PROOF. Assertion 4.3.1 follows from Theorem 2.2. For the proof of 4.3.2, it suffices to show that for any open sets U, $V \subset M$ there are a point $x \in U$ and an integer n such that $f^n x \in V$. By virtue of (4.1), one can select periodic points x_0, $x_1, \ldots, x_n$, with $x_0 \in U$ and $x_n \in V$, such that

$$W^u(x_k) \cap W^s(x_{k+1}) \neq \varnothing, \quad k = 0, \ldots, n-1.$$

Let N be the common period of these points. Then, replacing f by f^N, one may suppose that all the points x_k are fixed. Let $y_k \in W^u(x_k) \cap W^s(x_{k+1})$. It suffices to prove that $W^u(x_0) \cap V \neq \varnothing$, since a sufficiently iterated inverse image of a point in this intersection will lie in U. We shall prove, by induction on k, that x_k belongs to the closure $\overline{W^s(x_0)}$ of $W^s(x_0)$. This is obvious for $k = 1$, since $W^s(x_0) \in f^{nN} y_0 \to x_1$. Thus, assume that $x_k \in \overline{W^s(x_0)}$. Since $f^{nN} y_k \in W^s(x_k)$, we have $x_{k+1} \in \overline{W^s(x_k)}$. Let $z_n \in W^s(x_0)$, with $z_n \to x_k$ as $n \to \infty$. By assertion P.4 of the theorem on stable manifolds, any compact part of $W^s(x_k)$ can be approximated, with arbitrary precision in the C^1-topology, by a compact part of $W^s(z_n)$ for sufficiently large n. Therefore $W^s(x_0) = \bigcup_n W^s(z_n) \supset \overline{W^s(x_k)} \ni x_{k+1}$. Thus 4.3.2 is proved. At the same time, we have proved 4.3.3 for periodic points. Now let $x \in M$ be an arbitrary point, V an arbitrary open subset of M, and y a periodic point in V. Choose an ϵ such that $W^u_\epsilon(y) \subset V$ and an r such that the set $f^r(W^u_\epsilon(y))$ intersects any ball of radius $\epsilon_2(\epsilon_1)$, where ϵ_1 is chosen as in Lemma 3.1, and $\epsilon_2(\delta)$ is the function in Lemma 3.2. Then, in particular, $f^r(W^u_\epsilon(y)) \cap W^s_{\epsilon_1}(f^r(x)) \neq \varnothing$; therefore $W^u_\epsilon(y)$ (and hence also V) intersects $f^{-r}(W^s_\epsilon(f^r x)) \subset W^s(x)$. This completes the proof of the theorem.

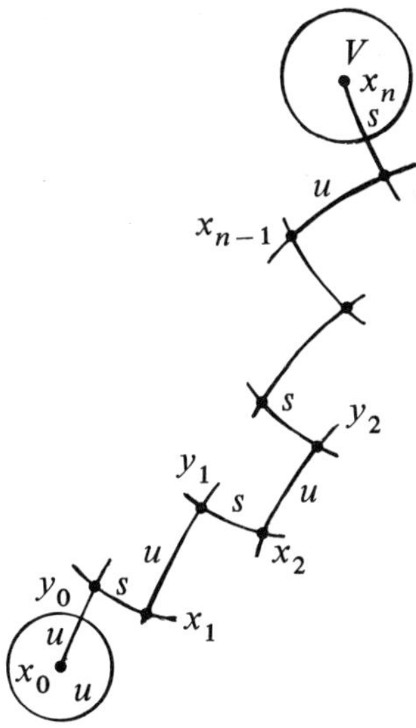

FIGURE 6

REMARK 4.1. We note that each of the properties (4.1), 4.3.1, 4.3.2, and 4.3.3 implies all the others. In fact, (4.1) clearly follows from 4.3.1 or 4.3.2. To deduce (4.1) from 4.3.3, we choose a periodic point $x_0 \in M$, which exists by Theorem 2.2. Then each of the manifolds $W^s(x_0)$ and $W^u(x_0)$ is everywhere dense, and, by virtue of their transversality, the intersection $W^s(x_0) \cap W^u(x_0)$ is also everywhere dense. But clearly any point of such an intersection is nonwandering.

EXERCISE 4.1. Prove that an Anosov diffeomorphism satisfying (4.1) has the following *domain mixing property*, stronger than topological transitivity: for any open subsets U, $V \subset M$, there is an N such that $f^n U \cap V \neq \varnothing$ for all $n \geqslant N$.

All three assertions of Theorem 4.3 may be strengthened by relating them to a certain measure on *M*.

THEOREM 4.4. *Let P_n be the set of points of all the periodic trajectories with period not exceeding n, and let $N(P_n)$ be the number of elements of P_n. Let μ_n denote the measure which assigns equal values $1/N(P_n)$ to each point of P_n and is zero outside P_n. Then the following assertions are true*:

4.4.1. *The sequence of measures μ_n converges weakly to a measure μ which is invariant with respect to f and positive on any open subset of M.*

4.4.2. *f is ergodic with respect to the measure μ.*

4.4.3. *The foliations W^s and W^u are metrically transitive with respect to μ; that is, any measurable set consisting of entire leaves of W^s or W^u either has measure* 0 *or is complementary to a set of measure* 0.

These assertions were proved by Sinaĭ [15] and Margulis [6]. The measure μ has a series of other remarkable properties, which will not be explicitly described here. We note that many of the deepest results on Anosov diffeomorphisms (and Anosov flows) obtained up to the present time are related to metrical questions (see [1], [17] and [2]).

4. The topological theory of Anosov diffeomorphisms is still far from complete. This theory is confronted by the following three fundamental questions:

1) For every Anosov diffeomorphism $f\colon M \longrightarrow M$, is $\Omega(f) = M$?

2) On which manifolds do Anosov diffeomorphisms exist?

3) When are two Anosov diffeomorphisms, on the same manifold *M*, topologically conjugate?

The statement of the first question is quite concrete. Despite its apparently technical character, it seems to us that the importance of this question is fully and convincingly demonstrated by Theorems 4.3 and 4.4 as well as all the rich metrical theory of Anosov diffeomorphisms which has been constructed on the basis of the assumption that $\Omega(f) = M$.

Moreover, as existing results show, an affirmative answer to this question would serve as a necessary foundation for serious progress toward the solution of problems 2) and 3).

The following theorem constitutes the only serious achievement in this direction.

THEOREM 4.5 (NEWHOUSE [19]). *If* $\dim E^s_x = 1$ (*or* $\dim E^u_x = 1$), *then* $\Omega(f) = M$.

The second and third questions require more concrete formulation. Probably, in the first of these questions, it would be natural to determine reasonable necessary (and, in the latter, sufficient) conditions. Here are two plausible conjectures:

CONJECTURE 4.1. If there exists an Anosov diffeomorphism on the manifold *M*, then the universal covering of *M* is a euclidean space.

CONJECTURE 4.2. If f_1 and f_2 are Anosov diffeomorphisms on the manifold M and if the induced automorphisms f_{1*} and f_{2*} of the fundamental group $\Pi_1(M)$ are conjugate, then f_1 and f_2 are topologically conjugate.

Anosov diffeomorphisms for which $\dim E_x^s = 1$ (or $\dim E_x^u = 1$) are said to be of *codimension one.*

THEOREM 4.6 (FRANKS AND NEWHOUSE). *Any Anosov diffeomorphism of codimension one is topologically conjugate to an algebraic automorphism of the torus (see Example* 1.3).

This theorem was proved by Franks [3] under the assumption that $\Omega(f) = M$. Combined with the theorem of Newhouse (Theorem 4.5), this result of Franks provides the best possible statement in the case under consideration.

COROLLARY 4.1. *Any Anosov diffeomorphism on a manifold of dimension* 2 *or* 3 *is topologically conjugate to an algebraic automorphism of the torus* (*that is,* $M = \mathbf{T}^2$ *or* $\mathbf{T}^3$).

The assertion of Corollary 4.1 for dimension 3 probably cannot be proved much more easily than the general assertion of Theorem 4.6. As regards the assertion for dimension 2, this is a simple fact which can be obtained from completely elementary considerations, although such a proof has apparently not been published.

Franks has also obtained several other results directed toward the topological classification of Anosov diffeomorphisms. The following theorem, proved in [3] and [20], contains Conjecture 4.2 in one special case.

THEOREM 4.7. *If f is an Anosov diffeomorphism of the n-dimensional torus* $\mathbf{T}^n$, $\Omega(f) = \mathbf{T}^n$, f_* *is the induced transformation of* $H_1(\mathbf{T}^n, \mathbf{R})$ *and* (*spec* f_*) $\cap \Delta = \varnothing$, *then f is topologically conjugate to an algebraic automorphism.*

Another result of Franks in [3] concerns a certain class of diffeomorphisms (Π_1-diffeomorphisms), introduced by him, for which he proves Conjectures 4.1 and 4.2 under certain assumptions regarding the fundamental group $\Pi_1(M)$. Although the equivalence of Anosov diffeomorphisms and Π_1-diffeomorphisms has not been formally established, there exist no counterexamples.

§5. Locally maximal hyperbolic sets

1. Arbitrary hyperbolic sets constitute an immense class of objects. They include, for example, all shift invariant closed subsets of the space of dyadic sequences (see Example 1.5). Various attempts have been made to single out, from the totality of hyperbolic sets, a class which is sufficiently wide but permits, to some extent, an overall survey. One of the most successful of these attemps led to the concept of local maximality (Anosov [9], Alekseev [21]).

DEFINITION 5.1. A hyperbolic set Λ of the diffeomorphism $f\colon U \longrightarrow M$ is said to be *locally maximal* if there is an open neighborhood $U(\Lambda) \supset \Lambda$ such that any invariant set Λ', $\Lambda \subset \Lambda' \subset U(\Lambda)$, coincides with Λ.

As will be shown below, many of the assertions of §4, relating to Anosov diffeomorphisms, can, with certain modifications, be carried over to locally maximal hyperbolic sets.

2. EXAMPLE 5.1. Any finite hyperbolic set is locally maximal.

EXAMPLE 5.2. Let $V \subset U$ be an open set such that $\Lambda = \bigcap_{-\infty}^{+\infty} f^n V$ is closed and hyperbolic. Then Λ is locally maximal. From this, in particular, follows the local maximality of the Cantor set in Example 1.5 and that of the solenoid in Example 1.6.

EXAMPLE 5.3. A set Λ consisting of a hyperbolic fixed point x_0 and the trajectory of a homoclinic point $x \in W^s(x_0) \cap W^u(x_0)$ is not locally maximal.

This is easily seen in the case of Example 1.5, considering the fixed point $x_{00\ldots}$ and the homoclinic point $x_{\ldots 00100\ldots}$. The trajectory of the latter consists of the points x_ω, where ω contains a single 1 and otherwise zeros. Fix N and add to this set the trajectory of the point $x_{\omega'}$, where ω' contains two entries equal to 1, separated by N zeros, and the remaining entries of ω' are zero.

In the general case, one must choose N so that the points $f^N x$ and $f^{-N} x$ lie in a sufficiently small neighborhoood of x_0 and $W^u_{\epsilon_1}(f^N x)$ intersects $W^s_{\epsilon_1}(f^{-N} x)$ in a unique point x'. Clearly, the set $\Lambda \cup (\bigcup_n f^n x')$ is contained in the prescribed neighborhood of Λ for sufficiently large N (see Figure 7).

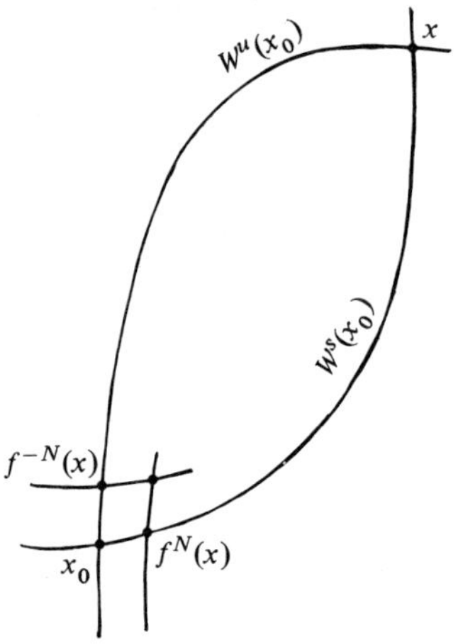

FIGURE 7

We return to Example 1.5.

Given any neighborhood of Λ, one can choose N so that the trajectory of the point $x_{\omega'}$ lies entirely in this neighborhood. But any preassigned neighborhood of Λ already contains a locally maximal set (for example, the set of points of the form x_ω, where ω ranges over all possible dyadic sequences having the property that between any two adjacent entries equal to 1 there is a block of not less than N zeros, and N is sufficiently large).

This example shows that, in some cases, a hyperbolic set which is not locally maximal can be approximated by locally maximal hyperbolic sets.

PROBLEM. Let Λ be a hyperbolic set of the diffeomorphism $f\colon U \longrightarrow M$, and V an open neighborhood of Λ. Does there exist a locally maximal hyperbolic set Λ' such that $V \supset \Lambda' \supset \Lambda$?

One might weaken the assertion and merely require the existence of some (not necessarily contained in V) locally maximal hyperbolic set $\Lambda' \supset \Lambda$. However, these are apparently equivalent questions, since one may assume that the diffeomorphism f is "not hyperbolic" outside V. The following simplification of the problem appears to be more substantial.

Let Λ_0 be a locally maximal hyperbolic set of f, Λ a closed subset of Λ_0, and V a neighborhood of Λ. Does there exist a locally maximal hyperbolic set Λ' such that $V \supset \Lambda' \supset \Lambda$? An affirmative answer to this question seems very plausible, but an explicit proof is lacking, even for the case of Anosov diffeomorphisms.

3. DEFINITION 5.2 (SMALE [22]). A hyperbolic set Λ of the diffeomorphism $f\colon U \longrightarrow M$ is said to have a *local product structure* if there is a $\delta > 0$ such that from $x_1, x_2 \in \Lambda$ and $W^s_\delta(x_1) \cap W^u_\delta(x_2) \neq \varnothing$ it follows that

$$r(x_1, x_2) = W^s_\delta(x_1) \cap W^u_\delta(x_2) \in \Lambda.$$

REMARK. One may assume that $\delta < \epsilon_1$, where ϵ_1 comes from Lemma 3.1, so that $W^s_\delta(x_1) \cap W^u_\delta(x_2)$ consists of no more than one point. From Definition 5.2, the theorem on stable manifolds, and the lemmas of §3 it is not difficult to deduce that a hyperbolic set Λ having a local product structure is representable as a canonical direct product

$$(\Lambda \cap W^s_\epsilon(x)) \times (\Lambda \cap W^u_\epsilon(x))$$

in a neighborhood of each point $x \in \Lambda$. A precise formulation of this property is easily obtained by analogy with Theorem 4.1 (see [19]).

Of still greater interest is the following theorem, proved by Hirsch, Palis, Pugh and Shub [23] in an explicit formulation somewhat different from ours.

THEOREM 5.1. *For a hyperbolic set Λ of a diffeomorphism $f\colon U \longrightarrow M$, the following properties are equivalent.*

5.1.1. *Λ has a local product structure.*

5.1.2. *Λ is locally maximal.*

5.1.3. *There exists an open neighborhood $U(\Lambda) \supset \Lambda$ such that, given any point y for which $f^n y \in U(\Lambda)$ for all $n \geq 0$, there is a point $x \in \Lambda$ for which $y \in W^s(x)$.*

PROOF. We first prove the equivalence of 5.1.1 and 5.1.2.

The local product structure (5.1.1) follows easily from the local maximality (5.1.2). In fact, let $\delta > 0$ ($\delta < \epsilon_1$) be such that

$$\bigcup_{x \in \Lambda} (W^s_\delta(x) \cup W^u_\delta(x)) \subset U(\Lambda).$$

Suppose $W^s_\delta(x_1) \cap W^u_\delta(x_2) \neq \varnothing$. The set $\Lambda' = \Lambda \cup (\bigcup_n f^n r(x_1, x_2))$ is closed, invariant with respect to f, and contained in $U(\Lambda)$. By 5.1.2, $\Lambda' \subset \Lambda$ and, consequently, $r(x_1, x_2) \in \Lambda$.

In order to deduce 5.1.2 from 5.1.1, we first prove the following lemma.

LEMMA 5.1. *Let Λ be a hyperbolic set having a local product structure. There exist numbers $\delta_1, \delta_2 > 0$ such that if $x \in \Lambda$, $y \notin \Lambda$ and $y \in W^u_{\delta_1}(x)$, then, for some $N > 0$,*

$$\rho(f^N y, \Lambda) > \delta_2.$$

PROOF. For $y \in W^u_{\delta_1}(x)$ (where δ_1 will be chosen below), we write

$$a_n = \min_{z \in \Lambda \cap W^s(f^n x)} \rho^u(f^n y, z).$$

(Recall that ρ^u is the metric on the unstable manifold.) If $\delta_1 < \epsilon_4$, where ϵ_4 is the number in Lemma 3.4 defined here for f^{-1}, then there are numbers $K, \mu > 1$, independent of n, x and y, such that $\mu a_n \leqslant a_{n+1} \leqslant K a_n$. Let $\delta_1 < ½$ be a number satisfying the conditions of Definition 5.2 and, in addition, so small that if $x_1, x_2 \in \Lambda$, $x'_1 \in W^u_{\delta_1}(x_1)$ and $\rho(x'_1, x_2) < \delta_1^2$, then $W^s_{2\delta_1}(x_2) \cap W^u_{2\delta_1}(x_1) \neq \varnothing$ and

$$\rho^u(x'_1, r(x_2, x_1)) < \frac{\delta_1}{K}$$

(see Lemma 3.5). Choose $n \geqslant 0$ so that

$$\delta_1/K < a_n < \delta_1. \tag{5.1}$$

Clearly, $f^n y \in W^u_{\delta_1}(x')$, where $x' \in \Lambda$. We shall show that one may take $\delta_2 = 2\delta_1^2$. In fact, suppose that $z \in \Lambda$ and $\rho(f^n y, z) \leqslant \delta_2$. Then $\rho^u(f^n y, r(z, x')) < \delta_1/K$, by the choice of δ_1.

But, by virtue of 5.1.1, $r(z, x') \in \Lambda$ and consequently $a_n < \delta_1/K$, which contradicts (5.1) (see Figure 8). The lemma is proved.

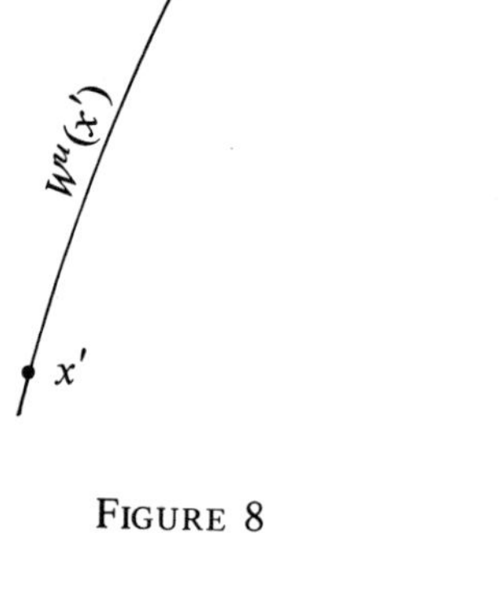

FIGURE 8

We now continue with the proof of Theorem 5.1. Let Λ' be a hyperbolic set containing all trajectories sufficiently close to Λ, and let $z \in \Lambda' \backslash \Lambda$, with distance $\rho(z, \Lambda) < \epsilon_2(\delta_2^2)$ (the function $\epsilon_2(\delta)$ of Lemma 3.2 is here constructed for the set Λ'). Then there is a point $x \in \Lambda$ for which $W^s_{\delta_2^2}(z) \cap W^u_{\delta_2^2}(x) \neq \varnothing$. Suppose that $r(z, x) \notin \Lambda$. By Lemma 5.1, for some $n \geqslant 0$ we have $\rho(f^n(r(z, x)), \Lambda) > \delta_2$, and therefore

$$\begin{aligned}\rho(f^n z, \Lambda) &> \delta_2 - \rho(f^n z, f^n(r(z, x)))\\ &\geqslant \delta_2 - \rho^s(f^n z, f^n(r(z, x))) \geqslant \delta_2 - \delta_2^2 > \delta_2/2.\end{aligned}$$

A similar inequality is obtained under the assumption that $r(x, z) \notin \Lambda$. Thus, if the trajectory of z lies in a δ_3-neighborhood of Λ, where $\delta_3 = \min(\delta_2/2, \epsilon_2(\delta_2^2))$, then $r(x, z) \in \Lambda$ and $r(z, x) \in \Lambda$ for some $x \in \Lambda$, where $\rho(x, z)$ is sufficiently small. By Lemma 5.1.1, $z = r(r(z, x), r(x, z)) \in \Lambda$. Thus, Λ is locally maximal.

It remains to deduce property 5.1.3 from local maximality. Thus, let V be some neighborhoood of Λ and let $f^n y \in V$, $n = 0, 1, 2, \ldots$. We shall show how one must choose V in order to ensure that $y \in W^s(x)$, where $x \in \Lambda$.

Consider a point $x' \in \Lambda$, sufficiently close to y. The sequence

$$x_n = \begin{cases} f^n y, & n \geqslant 0,\\ f^n x', & n < 0,\end{cases}$$

constitutes an ϵ-trajectory of f. If ϵ is sufficiently small, then, by the theorem on ϵ-trajectories, there is a point $x \in U$ such that $\rho(f^n x, x_n) < \delta$, where δ can be arbitrarily prescribed. In particular, δ can be fixed so that the entire trajectory $\{f^n x\}$ lies in the neighborhood $U(\Lambda)$ appearing in Definition 5.1, and therefore $x \in \Lambda$. Thus $\rho(f^n y, f^n x) < \delta$ for $n \geqslant 0$, but if δ satisfies the conditions of the theorem on stable manifolds, then, by assertion P.5 of that theorem, $y \in W^s_\delta(x)$. This completes the proof of Theorem 5.1.

4. Let Λ be a locally maximal hyperbolic set of the diffeomorphism f and let f' be a diffeomorphism sufficiently close to f in the C^1-topology. By Theorem 2.3, f' has an invariant hyperbolic set $\Lambda_{f'}$, close to Λ, and there exists a homeomorphism ψ: $\Lambda_{f'} \longrightarrow \Lambda$, close to the identity, such that $\psi \cdot f'|_{\Lambda_{f'}} = f|_\Lambda \cdot \psi$.

We shall prove that the property of local maximality is rough.

THEOREM 5.2. *If the distance between f and f' in the C^1-topology is sufficiently small, then the set $\Lambda_{f'}$ is locally maximal.*

PROOF. Suppose that $\Lambda_{f'}$ is not locally maximal. Then, by Theorem 5.1, $\Lambda_{f'}$ does not have a local product structure. Let $r(x_1, x_2) = W^s_\delta(x_1) \cap W^u_\delta(x_2) \notin \Lambda_{f'}$, where the stable and unstable manifolds refer to the diffeomorphism f'. Clearly,

$$\rho(f'^n x_1, f'^n r(x_1, x_2)) < \delta, \qquad \rho(f'^{-n} x_2, f'^{-n} r(x_1, x_2)) < \delta$$

for $n \geqslant 0$. If f' is sufficiently close to f, then for the point $q = \psi(r(x_1, x_2))$ we have

$$\rho(f^n\psi(x_1), f^n q) < 2\delta, \qquad \rho(f^{-n}\psi(x_2), f^{-n}q) < 2\delta \tag{5.2}$$

for $n \geqslant 0$.

If δ is sufficiently small, then sufficiently small local stable manifolds of f, passing through $\psi(x_1)$ and $\psi(x_2)$, intersect in a point which for brevity we denote by q'. Since the set Λ is locally maximal, and therefore has a local product structure, it follows that $q' \in \Lambda$. For $n \geqslant 0$ we have

$$\rho(f^n\psi(x_1), f^n q') < \theta, \quad \rho(f^{-n}\psi(x_2), f^{-n}q') < \theta, \tag{5.3}$$

where the number θ can be made sufficiently small if δ is small and f' is sufficiently close to f. Combining (5.2) and (5.3), and applying Theorem 2.4, we obtain $q' = q$; that is, $\Lambda_{f'} \ni \psi^{-1}(q') = r(x_1, x_2)$. This contradiction proves the lemma.

Let $U(\Lambda)$ be the neighborhood appearing in Definition 5.1, $f\colon\ U(\Lambda) \longrightarrow M$, and $n \geqslant 0$.

We shall use the notation

$$W_\epsilon^{s,n}(x) = f^{-n}W_\epsilon^s(f^n x),$$

$$W^s(x) = \bigcup_{n\geqslant 0} W_\epsilon^{s,n}(x),$$

$$W_\epsilon^{u,n}(x) = f^n W_\epsilon^u(f^{-n}x),$$

$$W^u(x) = \bigcup_{n\geqslant 0} W_\epsilon^{u,n}(x).$$

THEOREM 5.3. *Let Λ be a locally maximal hyperbolic set of the diffeomorphism $f\colon\ U(\Lambda) \longrightarrow M$. There is a natural number M such that the set $\Delta' = \Omega(f|_\Lambda)$ is representable as a union of a finite number of pairwise disjoint subsets $\Lambda_1, \ldots, \Lambda_k$ which are invariant with respect to f^M, and such that the following conditions hold:*

5.3.1. *The diffeomorphism f^M is topologically transitive on each Λ_i, $i = 1, \ldots, k$.*

5.3.2. *The points x_1 and x_2 belong to one and the same set Λ_i if and only if $W^s(x_1) \cap W^u(x_2) \neq \varnothing$, $W^u(x_1) \cap W^s(x_2) \neq \varnothing$, and both intersections are transversal in at least one point.*

5.3.3. $\overline{W^s(x)} \supset \Lambda_i$ *for any point $x \in \Lambda_i$.*

PROOF. By Theorem 2.2, the closure of the set of periodic points of f contains Λ'. By the local maximality of Λ, any periodic point of f lying in some neighborhood of Λ belongs to Λ and therefore to Λ'. Thus Λ' contains an everywhere dense subset P consisting of periodic points.

We introduce the following equivalence relation in the set P. Let $x_1, x_2 \in P$; then $x_1 \sim x_2$ if

$$W^s(x_1) \cap W^u(x_2) \neq \varnothing, \quad W^u(x_1) \cap W^s(x_2) \neq \varnothing,$$

and each of these intersections is transversal in at least one point. The reflexivity and symmetry of this relation are obvious. We verify transitivity. Let $x_1, x_2, x_3 \in P$, $x_1 \sim x_2$ and $x_2 \sim x_3$; let y_1 be a point of transversal intersection of $W^u(x_1)$ and $W^s(x_2)$, and y_2 a point of transversal intersection of $W^u(x_2)$ and $W^s(x_3)$. Let N be a common period of the points x_1, x_2 and x_3. Clearly, $W_\epsilon^{u,m}(f^{kN}y_1) \subset W^u(x_1)$ for any k and m. But, as $k \longrightarrow \infty$, $f^{kN}y_1 \longrightarrow x_2$, and therefore the manifold $W_\epsilon^{u,m}(f^{kN}y_1)$ approaches $W_\epsilon^{u,m}(x_2)$, and, for appropriate m, n and sufficiently large k, intersects $W_\epsilon^{s,n}(x_3)$ transversally in a point sufficiently close to y_2. Similarly, one proves that $W^s(x_1) \cap W^u(x_3)$ is nonempty and contains a point of transversal intersection.

By Lemma 3.2, there is an $\epsilon > 0$ such that if $x_1, x_2 \in P$ and $f(x_1, x_2) < \epsilon$, then $x_1 \sim x_2$, since in this case the local stable and unstable manifolds intersect. Covering the set Λ with a finite number of balls of radius ϵ, we deduce that there can only be a finite number of equivalence classes, which we denote by $P_1, \ldots, P_k$. Since the sets P_i are transformed into one another under the action of f, there is an $M > 0$ such that $f^M P_i = P_i$, $i = 1, \ldots, k$. Let Λ_i denote the closure of the set of points belonging to the class P_i, $i = 1, \ldots, k$. Clearly, the sets Λ_i are pairwise disjoint and $\bigcup_1^k \Lambda_i = \Lambda'$. Assertion 5.3.1 has virtually been proved, since for any pair of open subsets $U_1, U_2 \subset \Lambda_i$ there are points $x_1 \in P \cap U_1$ and $x_2 \in P \cap U_2$, and because $x_1 \sim x_2$ there is a point y such that $f^{-kNM}y \longrightarrow x_1$ and $f^{kNM}y \longrightarrow x_2$ as $k \longrightarrow +\infty$, where N is a common period of the points x_1 and x_2. By the same argument, it also follows that $\overline{W^s(x)} \supset \Lambda_i$ for any point $x \in P \cap \Lambda_i$. Thus, assertions 5.3.2 and 5.3.3 are proved for periodic points.

Let $x_1, x_2 \in \Lambda$ and $y \in W^s(x_1) \cap W^u(x_2)$. Then $y \in \Lambda$, since the trajectory of y is entirely contained in $U(\Lambda)$. Therefore, at y, $W^s(x_1)$ is tangent to E_y^s, $W^u(x_2)$ is tangent to E_y^u, and consequently their intersection is transversal. Assertion 5.3.3 for an arbitrary point $x \in \Lambda_i$ can now be proved in exactly the same manner as the analogous assertion 4.3.3 for Anosov diffeomorphisms. From the density of stable and unstable manifolds in Λ_i, it follows at once that the intersections described in 5.3.2 are nonempty and transversal at all their points. Now suppose that $x_1, x_2 \in \Lambda_i$, and that these intersections are nonempty and transversal. Then there are periodic points x_1' and x_2', arbitrarily close to x_1 and x_2, respectively, for which these intersections are also nonempty and transversal, and therefore $x_1' \sim x_2'$. Hence x_1 and x_2 lie in the same set Λ_i.

REMARK 5.1. Assertion 5.3.2 defines an equivalence relation on Λ'.

REMARK 5.2. If periodic points are dense in Λ, then $\Lambda = \Lambda'$, and in this case Theorem 5.3 relates to the set Λ itself; in particular, the theorem applies to Anosov diffeomorphisms satisfying condition (4.1). Since the decomposition $\Lambda' = \Lambda_1 \cup \cdots \cup \Lambda_k$ can only consist of a single element if Λ' is connected, Theorem 4.3 on Anosov diffeomorphism follows from Theorem 5.3.

LEMMA 5.2. *Let x_1 and x_2 be (not necessarily distinct) hyperbolic periodic points of the diffeomorphism f, y_1 a transversal point of the intersection of $W^s(x_1)$ and $W^u(x_2)$, y_2 a transversal point of the intersection of $W^u(x_1)$ and $W^s(x_2)$, V a neighborhood of y_1, Λ the union of the trajectories of the points x_1, x_2, y_1, y_2, and V' a neighborhood of Λ. Then there is a periodic point $y \in V$ whose trajectory lies entirely in V'.*

PROOF. Clearly, the set Λ is hyperbolic (see Examples 1.7 and 1.8). One may assume that the neighborhood V' is such that the theorem on families of ϵ-trajectories applies to the diffeomorphism f in V'.

Let N be a common period of the points x_1 and x_2. Fix k and consider the finite sequence of points

$$y_1, fy_1, \ldots, f^{kN-1}y_1, x_2, f^{-kN+1}y_2, \ldots, f^{-1}y_2, y_2, fy_2,$$
$$f^{kN-1}y_2, x_1, f^{-kN+1}y_1, \ldots, f^{-1}y_1.$$

Denote the lth term of this sequence by z_l, $l = 0, 1, \ldots, 4kN - 1$. Given any $\epsilon > 0$, one can choose a k such that $\rho(z_{l+1}, fz_l) < \epsilon$, $l = 0, 1, \ldots, 4kN - 1$, since $f^{kN}y_1 \longrightarrow x_2$, etc.

From here on, we argue as we did in the derivation of Theorem 2.1 from the theorem on families of ϵ-trajectories. The space $X = X_{4kN} = \{0, 1, \ldots, 4kN - 1\}$, the diffeomorphism $g(l) = l + 1 \pmod{4kN}$, and the mapping φ: $X \longrightarrow V$ defined by $\varphi(l) = z_l$ satisfy the conditions of the theorem on families of ϵ-trajectories, provided k is chosen sufficiently large. Therefore, there is a periodic point y whose trajectory lies arbitrarily close to the sequence $\{z_l\}$. The lemma is proved.

COROLLARY 5.3. *There is a periodic point of the diffeomorphism f in any neighborhood of a homoclinic point (Example 1.7).*

In §6, we shall also require a generalization of Lemma 5.2, the proof of which we leave to the reader:

EXERCISE 5.1. Let $x_0, x_1, \ldots, x_{n-1}, x_n$ (where $x_n = x_0$) be hyperbolic periodic points of the diffeomorphism f: $U \longrightarrow M$, y_k $(k = 0, 1, \ldots, n - 1)$ a transversal point of the intersection of $W^u(x_k)$ and $W^s(x_{k+1})$, V a neighborhood of y_0, and V' a neighborhood of the union of the trajectories of x_k and y_k, $k = 0, 1, \ldots, n - 1$. Prove that there is a periodic point $y \in V$ whose trajectory lies in V'.

EXERCISE 5.2. Prove assertion 4.3.2, concerning the topological transitivity of Anosov diffeomorphisms, using the theorem on families of ϵ-trajectories instead of the inductive construction.

By means of Lemma 5.2, we shall prove that the sets Λ_i in Theorem 5.3 have a certain additional property, stronger than local maximality.

COMPLEMENT TO THEOREM 5.3. *If $x_1, x_2 \in \Lambda_i$ and $y \in W^s(x_1) \cap W^u(x_2)$, then $y \in \Lambda_i$.*

PROOF. We observe that, by the definition of the "global" manifolds $W^s(x_1)$ and $W^u(x_2)$, the point y belongs to Λ. Therefore these manifolds intersect transversally at y, for otherwise y could not belong to a hyperbolic set. It suffices to prove that $y \in \Lambda'$, since it is clear that the closure of the trajectory of y with respect to f^M lies in Λ_i, and consequently $y \notin \Lambda_j$ for $j \neq i$. Fix a neighborhood V of y, and approximate x_1 and x_2 by periodic points x'_1 and x'_2 such that $W^s(x'_1)$ and $W^u(x'_2)$ intersect at some point $y' \in V$. Since $x'_1, x'_2 \in \Lambda_i$, the manifold $W^u(x'_1)$ intersects $W^s(x'_2)$, and by Lemma 5.2 there are periodic points of f arbitrarily close to y'. Since the neighborhood V was arbitrarily chosen, y belongs to the closure of the set of periodic points of f, and consequently $y \in \Lambda'$.

COROLLARY 5.2. *The sets Λ' and $\Lambda_1, \ldots, \Lambda_k$ are locally maximal.*

PROOF. In the complement to Theorem 5.3, it was proved that $\Lambda_1, \ldots, \Lambda_k$ have a local product structure, and therefore, by Theorem 5.1, they are locally maximal. To prove the local maximality of Λ', it suffices to observe that, since the sets $\Lambda_1, \ldots, \Lambda_k$ are closed, sufficiently small local stable and unstable manifolds of points lying in different sets Λ_i cannot intersect.

Thus we have extended, to the case of locally maximal hyperbolic sets, certain assertions concerning Anosov diffeomorphisms which were proved in §4. Markov partitions and topological Markov chains serve as devices for a more detailed structural study of locally maximal hyperbolic sets. This method permits one to carry over, to locally maximal sets in which the periodic points are dense, a good many metrical (i.e., related to invariant measures) results concerning Anosov diffeomorphisms (e.g., generalizing the assertions of Theorem 4.4). (See Bowen's articles [24] and [25].)

§6. Axiom A, Ω-stability and structural stability

1. Up to this point, our considerations have, on the whole, been of a "local" character; we have studied the behavior of trajectories on a hyperbolic set itself, or in a small neighborhood of a hyperbolic set. However, one of the basic stimuli for the emergence of the theory of hyperbolic sets was the hope of delineating a qualitative picture of the trajectories of a "typical" diffeomorphism of a compact manifold. More precisely, we refer to the selection of a system of properties ("axioms") possessed by certain diffeomorphisms which form a dense open set in the space $\mathrm{Diff}^r(M)$ of all C^r-diffeomorphisms of an arbitrary compact manifold M. There is one case, which we have already considered, where the local and global viewpoints are contiguous, namely, that of Anosov diffeomorphisms. Since the Anosov diffeomorphisms form a nonempty open set in $\mathrm{Diff}^r(M)$ for certain manifolds M (see Examples 1.2–1.4 and Corollary 2.1), they ought to satisfy the system of axioms in question.

The efforts of a series of mathematicians working on the theory of smooth dynamical systems have for several years been concentrated around this problem (see [2], §I.6). Although the proposed system of axioms has not yet provided a solution to the problem, and, what is more, there has arisen a valid suspicion that a "good" system of axioms describing a dense open set of diffeomorphisms in $\mathrm{Diff}^r(M)$ may not exist, still, these efforts have not been expended in vain. In the first place, rather broad classes of diffeomorphisms have been selected, for which an interesting and substantial structural theory has been established. Second, these efforts have led to the formulation of plausible conjectures concerning necessary and sufficient conditions for structural stability and Ω-stability of diffeomorphisms, although it has turned out that these properties themselves are not typical ([26], [27]; see also Example 14 in §8 of Kušnirenko's lectures). Third, the examples of contravention of various "good" properties on an open set of diffeomorphisms (Smale [26], Abraham and Smale [27], Hirsch, Pugh and Shub [28], Simon [29], Newhouse [30]) have been constructed by using hyperbolic effects and, aside from their own immediate purpose, they have provided an abundant supply of complex situations which arise from the interaction of "poorly compatible" hyperbolic sets of a single diffeomorphism, and are important for general considerations, since they appear in a "rough" manner, i.e., on open sets in the space of diffeomorphisms.

2. The example of Anosov diffeomorphisms, as well as the stability of hyperbolic sets (Theorem 2.3), suggests the selection of a class of diffeomorphisms with more or less regular hyperbolic behavior.

AXIOM A (Smale). The set $\Omega(f)$ of nonwandering points of a diffeomorphism $f\colon M \longrightarrow M$ is hyperbolic, and the periodic points of f are dense in M.

PROPOSITION 6.1. *If a diffeomorphism f satisfies Axiom* A, *then $\Omega(f)$ is a locally maximal hyperbolic set.*

PROOF. It suffices to verify the local product structure property; furthermore, by the density of periodic points in $\Omega(f)$, it suffices to prove that $r(x_1, x_2) \in \Omega(f)$, where x_1 and x_2 are periodic points. But if the distance between x_1 and x_2 is sufficiently small (as compared with ϵ_1), the intersection $W^s_{\epsilon_1}(x_1) \cap W^u_{\epsilon_1}(x_2)$ is also nonempty (Lemma 3.2). By Lemma 5.2, the point $r(x_1, x_2)$ lies in the closure of the set of periodic points of f, and therefore $r(x_1, x_2) \in \Omega(f)$. The proposition is proved.

Applying Theorem 5.3 to this situation, we obtain the following theorem.

SMALE'S SPECTRAL DECOMPOSITION THEOREM.(3) *Let $f\colon M \longrightarrow M$ be a C^1-diffeomorphism satisfying Axiom* A. *Then the set $\Omega(f)$ is representable as the union of a finite number of pairwise disjoint closed invariant subsets $\Omega_1, \ldots, \Omega_k$, with the following properties*:

S.1. *f is topologically transitive on each Ω_i.*

S.2. *The points $x_1, x_2 \in \Omega(f)$ belong to the same set Ω_i if and only if $W^u(x_1) \cap W^s(x_2) \neq \varnothing$, $W^s(x_1) \cap W^u(x_2) \neq \varnothing$ and both intersections are transversal at at least one point.*

S.3. *$\overline{W^s(x)} \supset \Omega_i$ and $\overline{W^u(x)} \supset \Omega_i$ for any point $x \in \Omega_i$.*

(3) Although we deduce this theorem from the corresponding theorem (Theorem 5.3) for locally maximal hyperbolic sets, historically this all took place in a different order: Not long after the appearance of Smale's theorem [2], Anosov [9] observed that a similar assertion is also valid for locally maximal hyperbolic sets.

This theorem follows in a formal manner from Proposition 6.1 and Theorem 5.3, with the exception of one point, i.e. the sufficiency of the condition for x_1 and x_2 to belong to the same Ω_i (condition S.2), since Theorem 5.3 is concerned with global stable and unstable manifolds constructed in a small neighborhood of a set Λ, whereas in the spectral decomposition theorem they are constructed on the manifold M as a whole. The sufficiency of the condition follows easily from Lemma 5.2. In fact, a transversal point of intersection of $W^u(x_1)$ and $W^s(x_2)$ is nonwandering and consequently belongs, on the one hand, to the same set Ω_i as does x_1, and on the other hand, to the same set Ω_i as does x_2.

DEFINITION 6.1. The sets Ω_i in the spectral decomposition theorem are known as the *basic sets* of the diffeomorphism f.

PROPOSITION 6.2. $M = \bigcup_{x\in\Omega(f)} W^s(x)$.

PROOF. Let U be an arbitrary neighborhood of $\Omega(f)$. For each point $x \in M\setminus U$ there is a neighborhood $U(x)$ such that $U(x) \cap f^n U(x) = \varnothing$, $n = 1, 2, \ldots$. By virtue of its compactness, $M\setminus U$ can be covered by a finite number of neighborhoods like $U(x)$, which we denote by $V_1, \ldots, V_l$. From the construction of these sets it follows that, for any point $x \in M\setminus U$, $f^m x \notin M\setminus U$ for all $m \geqslant l$. Since $\Omega(f)$ is locally maximal, it follows from 5.1.3 that $f^m(x) \in W^s(y)$ for some $y \in \Omega_i$, provided only that U is chosen sufficiently small. Thus, $x \in W^s(y)$, and since U is arbitrary, the proposition is proved.

3. For some time, Smale presumed that the diffeomorphisms which do not satisfy Axiom A form a nowhere dense set in $\mathrm{Diff}^r(M)$. But then various examples appeared, showing that this is not the case (see [27]–[30]); we shall consider one of these examples. But in spite of this, Axiom A has proved to be a successful starting point for the formulation of conditions for the structural stability and Ω-stability of diffeomorphisms (for the definitions, see §§3 and 7 of Kušnirenko's lectures). We remark, for one thing, that Theorems 2.3 and 5.2 together with Proposition 6.1 allow one to prove that, for a small perturbation of a diffeomorphism satisfying Axiom A, the set of nonwandering points cannot diminish.

THEOREM 6.1. *Let f be a C^1-diffeomorphism satisfying Axiom* A. *For any $\delta > 0$ there is an $\epsilon > 0$ such that, if $f' \in \mathrm{Diff}^1(M)$ and $\rho_{C^1}(f, f') < \epsilon$, then the set $\Omega(f')$ contains a locally maximal subset $\Omega'(f')$, and there exists an imbedding ψ: $\Omega(f) \longrightarrow \Omega'(f')$ such that*

$$f'|_{\Omega'(f')} \circ \psi = \psi \circ f|_{\Omega(f)} \quad \text{and} \quad \max_{x\in\Omega(f)} \rho(x, \psi(x, \psi(x)) < \delta.$$

This is all that can be said about the structure of $\Omega(f')$ if one merely assumes that f satisfies Axiom A.

EXAMPLE 6.1 ("Ω-explosion" [22]). Let p, q and r be hyperbolic fixed points of a diffeomorphism f satisfying Axiom A, such that $\dim E^s_p = \dim E^s_q = \dim E^s_r$.

Let $p' \in W^u(p) \cap W^s(q)$, $q' \in W^u(q) \cap W^s(r)$ and $r' \in W^u(r) \cap W^s(p)$, where all three intersections are *not* transversal at the respective points p', q' and r'. Then, $p', q', r' \notin \Omega(f)$, since they cannot belong to a hyperbolic set (prove this!). We perturb the diffeomorphism in such a manner that the fixed points p, q and r are preserved, and so that transversal points of intersection of the appropriate manifolds appear in neighborhoods of the respective points p', q' and r'. These transversal points of intersection p'', q'' and r'' belong to the closure of the set of periodic points of the perturbed diffeomorphism (Exercise 5.1), and are therefore nonwandering. Thus, $p'', q'', r'' \in \Omega(f')$. But these points cannot belong to $\Omega'(f')$, since they are far from $\Omega(f)$. A concrete realization of this situation on the sphere S^2 is depicted in Figure 9. Here p, q and r are saddle points, the separatrices of which coincide in pairs, A, C and D are attractive points, and B and E are repulsive points. Thus, the sphere is partitioned into three regions (cells) I, II, III, and the topological configuration of the trajectories of f in each of these regions is uniquely determined. The set $\Omega(f)$ consists of eight fixed points. It can be proved (see Alekseev's lectures) that when transversal intersections appear in neighborhoods of p', q' and r', $\Omega(f')$ will contain a Cantor perfect set, upon which the action of f' is topologically transitive. The structure of $\Omega(f')$ is not clear. Also, it is not clear whether a diffeomorphism f' obtained by such a perturbation will satisfy Axiom A. However, by modifying this example, it can be proved that the diffeomorphisms satisfying Axiom A do not form an open set in $\mathrm{Diff}^r(S^2)$.

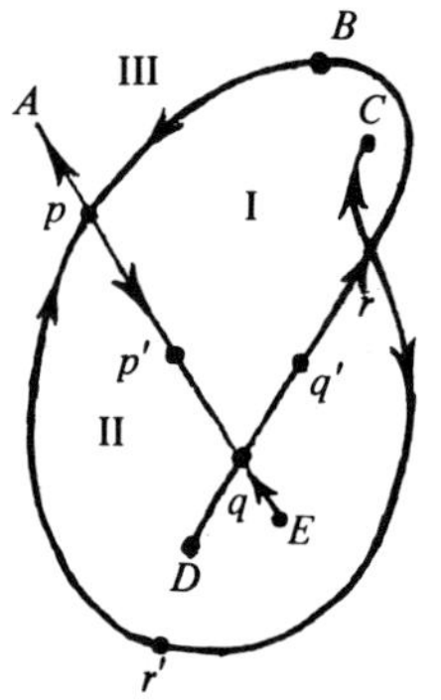

FIGURE 9

4. Thus, for a diffeomorphism to be Ω-stable, it must, in addition to Axiom A, satisfy some supplementary conditions. Let

$$W^s(\Omega_i) = \bigcup_{x \in \Omega_i} W^s(x).$$

From now on, instead of a decomposition satisfying conditions S.1–S.3, we shall consider an enlargement of this decomposition which is likewise invariant with respect to the diffeomorphism f. We shall also denote the elements of this decomposition (which are usually referred to as basic sets of f) by $\Omega_1, \ldots, \Omega_k$, which will cause no confusion, since the previous decomposition will receive no further mention. We introduce the following binary relation between basic sets of f: write $\Omega_i < \Omega_j$ if and only if $W^s(\Omega_i) \cap W^u(\Omega_j) \neq \varnothing$.

AXIOM ON ABSENCE OF CYCLES. There exists no chain of mutually distinct integers $i_1, \ldots, i_l$ $(l \geq 1)$ such that $\Omega_{i_1} < \Omega_{i_2} < \cdots < \Omega_{i_l} < \Omega_{i_1}$.

For the diffeomorphism f of Example 6.1, $\{p\} < \{q\} < \{r\} < \{p\}$.

EXERCISE 6.1. Let $i_1, \ldots, i_l, i_{l+1} = i_1$ be mutually distinct integers such that each intersection $W^s(\Omega_{i_r}) \cap W^u(\Omega_{i_{r+1}})$, $r = 1, \ldots, l$, is nonempty. Then at least one of these intersections is not transversal at any point.

The following theorem of Smale constitutes one of the basic achievements of the global theory relating to hyperbolic sets.

THEOREM ON Ω-SYABILITY. *If a C^1-diffeomorphism f: $M \longrightarrow M$ satisfies Axiom* A *and has the no-cycle property, then f is Ω-stable.*

Theorem 6.1, of course, constitutes the first step in the proof of the above theorem. The subsequent argument, excluding the possibility of a phenomenon of the "Ω-explosion" type, is based upon a construction known as filtration of f. The proof may be found in the original paper of Smale [22], or in Nitecki's book [14]. Smale has conjectured that Theorem 6.1 has a converse.

CONJECTURE 6.1. The diffeomorphism $f \in \mathrm{Diff}^r(M)$ is Ω-stable if and only if it satisfies Axiom A and has the no-cycle property.

Assuming that this conjecture is not refuted, the proof of necessity of the conditions for Ω-stability in Conjecture 6.1 probably constitutes a more difficult problem for $r \geq 2$ than for $r = 1$. Here one can draw a certain analogy with the so-called closing lemma (see §7.2 in Kušnirenko's lectures), which was proved by Pugh for $r = 1$ [31], whereas even an approach to the proof for $r \geq 2$ is lacking. At any rate, for $r = 1$ there has been serious progress toward a proof of the necessity of the conditions of Conjecture 6.1, using, incidentally, the closing lemma.

THEOREM 6.2 (FRANKS [7]). *For a diffeomorphism $f \in \mathrm{Diff}^1(M)$ to satisfy Axiom* A *and have the no-cycle property, it is necessary and sufficient that f possesses the following property*:

One can find a neighborhood N of f in $\mathrm{Diff}^1(M)$ *and a mapping φ: $N \longrightarrow C_0(\Omega(f), M)$ such that for $f' \in N$*

6.2.1. *$\varphi(f')$ is a homeomorphism, and* $\mathrm{Im}\,\varphi(f') = \Omega(f')$;

6.2.2. $\varphi(f') \circ f|_{\Omega(f)} = f'|_{\Omega(f)} \circ \varphi(f')$;

6.2.3. *$\Omega(f) = i$, the natural imbedding of $\Omega(f)$ in M*;

6.2.4. *φ is differentiable at f*; *and*

6.2.5. *there exists a constant $K > 0$ such that*

$$\sup_{x \in \Omega(f)} \rho(\varphi(f')x, x) \leq K \sup_{x \in M} (fx, f'x).$$

Conditions 6.2.1–6.2.5 amount to just Ω-stability, complemented by the two conditions 6.2.4 and 6.2.5.

The proof of Theorem 6.2 makes use of the definition of a hyperbolic set in terms of the operator f_* (see Theorem 1.1), and is based upon the fact that Ω-stability, together with the two supplementary properties, enables one to obtain certain estimates for this operator. We adduce one instructive lemma from this proof:

LEMMA 6.1. *If the diffeomorphism* $f \in \mathrm{Diff}^1(M)$ *is* Ω*-stable, then the operator* $Ig - f_*$ *is injective, where* $f_*\colon\ \Gamma(T_{\Omega(f)} M) \longrightarrow \Gamma(T_{\Omega(f)} M)$.

5. The open set of structurally unstable diffeomorphisms of the three-dimensional torus, which is constructed in Kušnirenko's lectures (§8, Example 14), consists of diffeomorphisms which satisfy the conditions of the Ω-stability theorem. Thus, we obtain an example of an open set of Ω-stable, but not structurally stable, diffeomorphisms. In this example, one exploits the fact that the property of tangency of manifolds which are defined in a topologically invariant manner can be given a topologically invariant meaning. The whole point is, of course, that certain manifolds $W^s(x)$ and $W^u(y)$ do not intersect transversally. We now give another example, of a simpler and nonrough nature, which illustrates certain phenomena.

EXAMPLE 6.2. Consider a torus $\mathbf{T}^2$, situated "vertically" in the space $\mathbf{R}^3$ with euclidean matrix (see Figure 10).

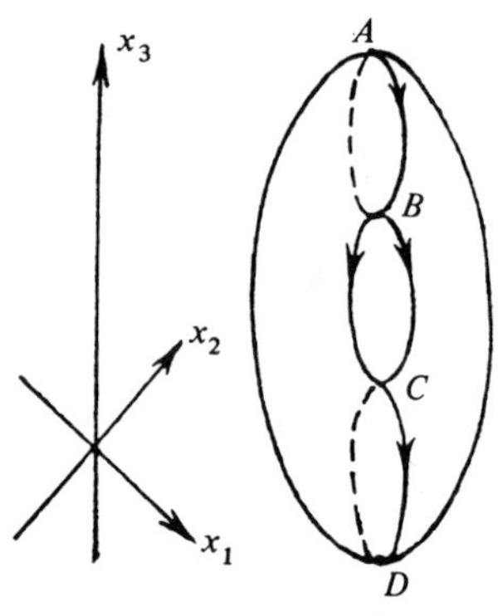

FIGURE 10

The flow defined by the gradient of the function x_3 on this torus has, as fixed points, the critical points A, B, C and D of the restriction of this function to the torus. A is a repulsive point, B and C are saddle points, and D is an attractive point. Consider the diffeomorphism f defined as the displacement along the trajectories of this flow in a unit time. The nonwandering points of this diffeomorphism are just A, B, C and D. Clearly, f satisfies the conditions of the Ω-stability theorem; moreover, $\{A\} > \{B\} > \{C\} > \{D\}$. At the same time there appears, in this example, a structurally unstable feature, namely the separatrix leading from the saddle point B to the saddle point C. Even while remaining in the class of displacements along the trajectories of smooth flows, one can perturb f in such a manner that the situation depicted in Figure 11 arises.

Here, the separatrices of B and C do not intersect. Therefore $\{A\} > \{B\} > \{D\}$ and $\{A\} > \{C\} > \{D\}$, but B and C are not connected by the relation $>$. Later on, we shall see (Theorem 6.3) that such a diffeomorphism is structurally stable. Another perturbation is also possible, also leading to structurally stable diffeomorphisms, which, however, can no longer be realized while remaining in the class of displacements along the trajectories of smooth flows (see Figure 12).

Here a heteroclinic point appears, but the set of nonwandering points consists, as before, of A, B, C and D. (Cf. Example 7 in §1 of Kušnirenko's lectures.)

The following conjecture was also formulated by Smale [32].

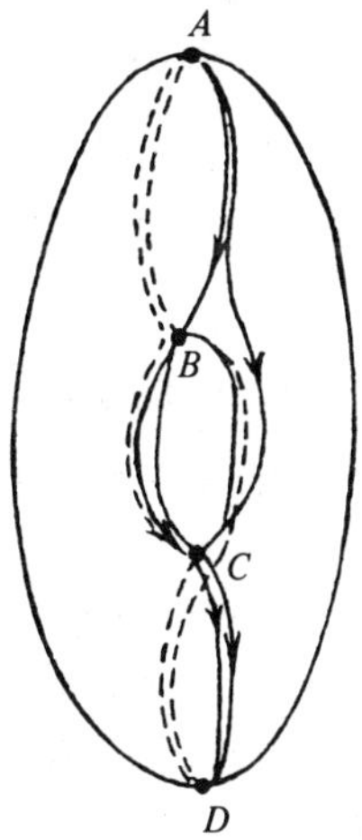

FIGURE 11

CONJECTURE 6.2. A diffeomorphism $f \in \mathrm{Diff}^r(M)$ is structurally stable if and only if it satisfies Axiom A and, for any points $x, y \in \Omega(f)$, the manifolds $W^s(x)$ and $W^u(y)$ intersect transversally.

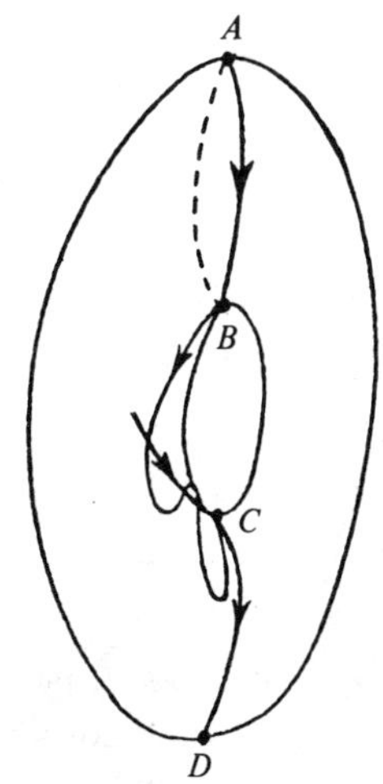

FIGURE 12

THEOREM 6.3 (ROBBIN [6]). *Suppose that $f \in \mathrm{Diff}^2(M)$ satisfies Axiom* A *and that, for any two points $x, y \in \Omega(f)$, the intersection $W^s(x) \cap W^u(y)$ is transversal at all its points. Then f is structurally stable in* $\mathrm{Diff}^1(M)$.

With the exception of a small gap (i.e., the theorem is proved for $f \in \mathrm{Diff}^2(M)$, and not for $f \in \mathrm{Diff}^1(M)$), this theorem subsumes all known results on structural stability.

If the set $\Omega(f)$ is finite, this gap can be eliminated.

THEOREM 6.4. (PALIS AND SMALE [33]). *If the diffeomorphism $f \in \mathrm{Diff}^1(M)$ is such that $\Omega(f)$ is a finite hyperbolic set and, for any two points $x, y \in \Omega(f)$, the manifolds $W^s(x)$ and $W^u(y)$ intersect transversally, then f is structurally stable.*

The key point in the proof of this theorem is the construction of a certain substitute for the stable and unstable manifolds, for all points of a certain neighborhood of $\Omega(f)$.

§7. Partially hyperbolic diffeomorphisms

1. In this section we shall consider a class of invariant sets of diffeomorphisms, wider than the class of hyperbolic sets.

DEFINITION 7.1. An invariant set Λ of the diffeomorphism $f\colon U \longrightarrow M$ is said to be *partially hyperbolic* if there are numbers α and β, $0 < \alpha \leqslant 1 \leqslant \beta$, such that $\operatorname{spec} f_* = S_1 \cup S_2 \cup S_3$, where $|\lambda| < \alpha$ for $\lambda \in S_1$, $\alpha < |\lambda| < \beta$ for $\lambda \in S_2$, $\beta < |\lambda|$ for $\lambda \in S_3$, and at least two of the sets S_1, S_2, S_3 are nonempty. Here, as usual, $f_*\colon \Gamma(T_\Lambda M) \longrightarrow \Gamma(T_\Lambda M)$.

PROPOSITION 7.1. *If Λ is a partially hyperbolic set of the diffeomorphism f, then there exists a continuous decomposition $T_\Lambda M = E^s \oplus E^0 \oplus E^u$, invariant with respect to Df, and such that*

$$\operatorname{spec}\left(f_*\big|_{\Gamma(E^s)}\right) = S_1, \quad \operatorname{spec}\left(f_*\big|_{\Gamma(E^0)}\right) = S_2, \quad \operatorname{spec}\left(f_*\big|_{\Gamma(E^u)}\right) = S_3.$$

This proposition is proved in two steps, in both of which one proceeds as in the proof of Theorem 1.1. Suppose that $S_1 \neq \varnothing$. We first consider the decomposition $\operatorname{spec} f_* = S_1 \cup S_2'$, where $S_2' = S_2 \cup S_3$, and prove Proposition 7.1 for this partition of the spectrum; this gives rise to a decomposition $T_\Lambda M = E^s \oplus E'$. If S_2 and S_3 are nonempty, we consider the operator f_* on $\Gamma(E')$ and, using the partition $\operatorname{spec}(f_*|_{\Gamma(E')}) = S_2 \cup S_3$, we thereby construct the decomposition $E' = E^0 \oplus E^u$.

The theorem on stable manifolds, with the exception of assertion P.5, carries over to partially hyperbolic sets. The proof of the theorem on stable manifolds, based upon successive approximations, also generalizes to partially hyperbolic sets.

DEFINITION 7.2. If the entire manifold M constitutes a partially hyperbolic set of a diffeomorphism $f \in \mathrm{Diff}^r(M)$, then f is called a *partially hyperbolic diffeomorphism*.

For partially hyperbolic diffeomorphisms, the distributions E^u and E^s are integrable, and their integral manifolds form foliations W^u and W^s. However, the sum of the dimensions of the leaves of these foliations will be less than dim M if $S_2 \neq \varnothing$.

PROPOSITION 7.2. *Let* $f\colon M \longrightarrow M$ *be a partially hyperbolic diffeomorphism. Then the following assertions are true*:

7.2.1. *There is a neighborhood* $\mathfrak{A}$ *of* f *in* Diff$^1(M)$ *such that any diffeomorphism* $f' \in \mathfrak{A}$ *is partially hyperbolic; moreover, for* f' *the numbers* α *and* β *in Definition* 7.1 *may be chosen to be the same as for* f.

7.2.2. *The subbundles* E^s, E^0 *and* E^u *of the tangent bundle TM depend continuously upon* f *in the* C^0-*topology.*

DEFINITION 7.3. The pair of foliations W^s, W^u induced by a partially hyperbolic diffeomorphism is said to be *transitive* if there exists an n such that for any two points $x, y \in M$ there is a sequence of points $x_0 = x, x_1, \ldots, x_{2n-1}, x_{2n} = y$ with

$$x_{2k+1} \in W^s(x_{2k}), \qquad x_{2k+2} \in W^u(x_{2k+1}), \qquad k = 0, \ldots, n-1.$$

The following theorem generalizes assertion 4.3.2 concerning Anosov diffeomorphisms.

THEOREM 7.1. *If* f *is a partially hyperbolic diffeomorphism,* $\Omega(f) = M$, *and the pair of foliations* W^s, W^u *is transitive, then* f *is topologically transitive.*

This theorem is proved under weaker assumptions in [38].

It is natural to suppose that the property of transitivity of the pair of foliations W^s, W^u is rough. However, at present this has been proved only under additional assumptions.

THEOREM 7.2 (BRIN AND PESIN [37]). *Let* $f \in$ Diff$^1(M)$ *be a partially hyperbolic diffeomorphism such that the distributions* E^s *and* E^u *are of class* C^∞, *and the pair of foliations* W^s, W^u *is transitive. Then there is a neighborhood* $\mathfrak{A}$ *of* f *in* Diff$^1(M)$ *such that for any diffeomorphism* $f' \in \mathfrak{A}$ *the pair of foliations* W^s, W^u *is transitive.*

EXAMPLE 7.1. Let φ_t be a geodesic flow on a surface of negative curvature (see [1] and Example 16 in Kušnirenko's lectures). The diffeomorphism $f = \varphi_1\colon M^3 \longrightarrow M^3$ is partially hyperbolic (see the definition of an Anosov flow in [1], §1). The distribution E^0 consists of the tangents to the flow φ_t; the distributions E^s and E^u are one-dimensional and commute with E^0. In Kušnirenko's lectures, $W^s(x)$ denotes a two-dimensional integral manifold of the distribution $E^s \oplus E^0$, but here we shall adhere to our former notation and denote by $W^s(x)$ a one-dimensional integral manifold of E^s. The pair of foliations W^s and W^u is transitive [1]. Consider the space Diff$^1(M^3, \mu)$ of diffeomorphisms of M^3 of class C^1 preserving the measure μ (a smooth measure, invariant with respect to the geodesic flow φ_t). Let $\mathfrak{A}' =$ Diff$^1(M^3, \mu) \cap \mathfrak{A}$, where $\mathfrak{A}$ is the neighborhood of f in Diff$^1(M^3)$ described in Theorem 7.2. Since $\Omega(g) = M^3$

for any $g \in \mathrm{Diff}^1(M^3, \mu)$, it follows by Theorems 7.1 and 7.2 that any diffeomorphism $f' \in \mathfrak{A}'$ is topologically transitive. On the other hand, by Theorem 4.6, f' is not an Anosov diffeomorphism, since the manifold M^3 is distinct from the torus. Thus, f' does not satisfy the conditions of Theorem 6.2 and is not Ω-stable (at least, not in the strong sense indicated in that theorem).

2. The question of integrability of the distribution E^0 is more complicated. For arbitrary partially hyperbolic sets, where the integral manifolds need not be entirely contained in the set Λ, nonuniqueness effects, for example, may appear. This in fact occurs in the case of a partially hyperbolic set consisting of a single point. Here is a very simple example.

Consider a neighborhood of the origin in the plane $\mathbf{R}^2$ and the diffeomorphism $f(x_1, x_2) = (x_1/2, x_2/4)$. Here $\Lambda = (0, 0)$, E^s is the x_1-axis, E^0 is the x_2-axis, and any parabola $x_2 = cx_1^2$ will serve as an integral manifold of E^0. In the case of partially hyperbolic diffeomorphisms where the integral manifolds of E^0 must fill up the compact manifold M, similar effects apparently cannot occur. Hirsch, Pugh and Shub [28] proved a theorem, which we state with certain simplifications:

THEOREM 7.3. *Let $f \in \mathrm{Diff}^1(M)$ be a partially hyperbolic diffeomorphism such that E^0 is integrable and satisfies one of the following two conditions*:

7.3.a. *E^0 satisfies a Lipschitz condition*;

7.3.b. *the space of integral manifolds of E^0 is Hausdorff.*

Then there is a neighborhood $\mathfrak{A}$ of f in $\mathrm{Diff}^1(M)$ such that, for any $f' \in \mathfrak{A}$, the corresponding distribution E^0 is integrable, and there exists a homeomorphism $g_{f'}: M \longrightarrow M$, depending continuously upon $f' \in \mathfrak{A}$, such that $g_f = \mathrm{Id}$ and $g_{f'}(W_f^0(x)) = W_{f'}^0(g_{f'} x)$. Here $W_f^0(x)$ and $W_{f'}^0(x)$ are the integral manifolds of the distributions E_f^0 and $E_{f'}^0$, respectively, containing the point x.

Thus, even though we are not immediately able to prove the integrability of E^0, nevertheless, the property of integrability is preserved under a small perturbation, at least if either of the conditions 7.3.a or 7.3.b is satisfied.

We shall use this theorem to construct an example of an open set in $\mathrm{Diff}^1(\mathbf{T}^4)$ consisting of topologically transitive but not Ω-stable diffeomorphisms. Also, we remark that, by means of this theorem, we can, without invoking Theorem 4.6, prove that the diffeomorphisms in Example 7.1 (and in similar but more general situations) are not Anosov diffeomorphisms.

EXAMPLE 7.2 (HIRSCH, PUGH AND SHUB [28]). Consider the diffeomorphism f of the four-dimensional torus $\mathbf{T}^4 = \mathbf{T}^2 \times \mathbf{T}^2$ defined by

$$f(\varphi, \psi) = (f_A^2 \varphi, f^\varphi \psi), \tag{7.1}$$

where $\varphi \in \mathbf{T}^2$, $\psi \in \mathbf{T}^2$, $A = \begin{pmatrix} 2 & 1 \\ 1 & 1 \end{pmatrix}$, f_A is the hyperbolic automorphism of $\mathbf{T}^2$ in Example 1.2, and f^φ is a diffeomorphism of $\mathbf{T}^2$ defined for each $\varphi \in \mathbf{T}^2$.

LEMMA 7.1. *Suppose that**

$$\max_{\varphi\in\mathbf{T}^2}\left(\max_{\psi\in\mathbf{T}^2}\|D(f^{\varphi})_{\psi}\|\right)<\lambda^2,$$

$$\max_{\varphi\in\mathbf{T}^2}\left(\max_{\psi\in\mathbf{T}^2}\|D(f^{\varphi})_{\psi}^{-1}\|\right)<\lambda^2$$

(where λ is the larger eigenvalue of the matrix A). Then f is a partially hyperbolic diffeomorphism; moreover,

$$(E_f^0)_{(\varphi,\psi)} = T_{(\varphi,\psi)}(\{\varphi\}\times\mathbf{T}^2),$$

while E_f^s and E_f^u are the one-dimensional distributions which map onto the corresponding distributions for f_A^2 under the projection $\pi_1\colon \mathbf{T}^4\to\mathbf{T}^2$, $\pi_1(\varphi,\psi)=\varphi$.

Since E_f^0 satisfies either one of the conditions 7.3.a or 7.3.b, we may apply Theorem 7.3. Suppose that the diffeomorphism f' lies in a sufficiently small C^1-neighborhood of f. Then $E_{f'}^0$ is C^0-close to E_f^0, and therefore each integral manifold $W_{f'}^0(\varphi, 0)$ is represented by a torus which is sufficiently close to the torus $\{\varphi\}\times\mathbf{T}^2$ in the C^1-topology. Let

$$W_{f'}^0(\varphi, 0)\cap(\mathbf{T}^2\times\{\psi\}) = (h(\varphi,\psi),\psi).$$

The mapping $g(\varphi,\psi)=(h(\varphi,\psi),\psi)$ is a homeomorphism of $\mathbf{T}^4$ onto itself which is sufficiently close to the identity. Clearly, $g^{-1}f'g(\varphi,\psi)=(\tilde{f}\varphi,\tilde{f}^{\varphi}\psi)$, where $\tilde{f}$ is a homeomorphism of $\mathbf{T}^2$ which is C^0-close to f_A^2, and, for each $\varphi\in\mathbf{T}^2$, $\tilde{f}^{\varphi}$ is a diffeomorphism which is close to f^{φ} in the C^1-topology, and depends continuously (but not necessarily smoothly) upon φ. We shall show that $\tilde{f}$ is topologically conjugate to f_A^2. In the theorem on families of ϵ-trajectories, we let $M=\mathbf{T}^2$, $f=f_A^2$, $X=\mathbf{T}^2$, $g=\tilde{f}$, and $\varphi=$ the identity mapping. Thus, there is a continuous mapping $\chi\colon \mathbf{T}^2\to\mathbf{T}^2$, close to the identity, such that $\chi\circ\tilde{f}=f_A^2\circ\chi$. It only remains to prove that χ is a homeomorphism. But this follows from the fact that $\tilde{f}$ has the property of expansiveness or dispersion of trajectories. We write

$$f''=(\chi\times\mathrm{id})g^{-1}f'g(\chi\times\mathrm{id})^{-1}.$$

Clearly,

$$f''(\varphi,\psi)=(f_A^2\varphi,\tilde{\tilde{f}}^{\varphi}\psi). \tag{7.2}$$

We shall sketch the proof of the following assertion.

PROPOSITION 7.3. *If in (7.2) the diffeomorphism $\tilde{\tilde{f}}^0$ is sufficiently C^1-close to f_A, then the homeomorphism f'' is topologically transitive.*

The crux of the proof of this proposition lies in the following lemma.

---

* *Translator's note.* The following inequalities were garbled in the original. I have changed them to what appears to be the intended form.

LEMMA 7.2. *Under the hypotheses of Proposition* 7.3, *there exists a neighborhood U of zero on the torus $\mathbf{T}^2$, with the following property*:

For any $\epsilon > 0$ there is a k such that $\pi_2(f'')^k(B)$ (where $\pi_2\colon\ \mathbf{T}^4 \longrightarrow \mathbf{T}^2$, $\pi_2(\varphi, \psi) = \psi$, and B is the ball of radius ϵ in $\{\varphi\} \times \mathbf{T}^2$) constitutes an ϵ-net on $\mathbf{T}^2$.

This lemma essentially deals with the behavior of the trajectories of a sequence of diffeomorphisms $f_n\colon\ \mathbf{T}^2 \longrightarrow \mathbf{T}^2$, where the f_n are uniformly C^1-close to f_A.

To prove Lemma 7.2, one must either use the theorem on constantly acting perturbations (with certain additional assertions) or prove, for this case, a particular nonautonomous variant of the theorem on families of ϵ-trajectories.

We shall show how to complete the proof of topological transitivity of f'' by the use of this lemma. Let U_1, V_1, U_2 and V_2 be balls in $\mathbf{T}^2$. We shall prove that there is a trajectory of f'' which intersects $U_1 \times V_1$ and $U_2 \times V_2$. Let $\varphi_1 \in W^s(0) \cap U_1$, $\varphi_2 \in W^u(0) \cap U_2$, $f_A^{2l}\varphi_1 \in U$ and $f_A^{-2l}\varphi_2 \in U$, where $W^s(0)$ and $W^u(0)$ are stable and unstable manifolds relative to the diffeomorphism f_A^2 and U is the neighborhood described in Lemma 7.2. Let B_1 and B_2 be balls in $(f_A^{2l}\varphi_1 \times \mathbf{T}^2) \cap ((f'')^l(U_1 \times V_1))$ and $(f_A^{-2l}\varphi_2 \times \mathbf{T}^2) \cap ((f'')^{-l}(U_2 \times V_2))$, respectively. For any sufficiently large k, one can find a point $\varphi^k \in \mathbf{T}^2$ such that

$$\varphi^k \in W^u_\delta(f_A^{2l}\varphi_1), \qquad f_A^{2k}\varphi^k \in W^s_\delta(f_A^{-2l}\varphi_2)$$

(where δ is a small fixed number) and

$$\rho(\varphi^k, f_A^{2l}\varphi_1) \longrightarrow 0, \qquad \rho(f_A^{2k}\varphi^k, f_A^{-2k}\varphi_2) \longrightarrow 0$$

as $k \longrightarrow \infty$.

Applying Lemma 7.2, we deduce that for sufficiently large k the intersection

$$(f'')^k(\{\varphi^k\} \times B_1) \cap (\{f_A^{2k}\varphi^k\} \times B_2)$$

is nonempty; and since, for sufficiently large k,

$$\{\varphi^k\} \times B_1 \subset (f'')^l(U_1 \times V_1), \quad \{f_A^{2k}\varphi^k\} \times B_2 \subset (f'')^{-l}(U_2 \times V_2),$$

we conclude that

$$(f'')^{k+2l}(U_1 \times V_1) \cap (U_2 \times V_2) \neq \varnothing.$$

Proposition 7.3 implies the topological transitivity of any diffeomorphism f' sufficiently close to a diffeomorphism (7.1) for which $f^0 = f_A$. Among such diffeomorphisms are structurally stable ones, e.g., the Anosov diffeomorphism $f_A^2 \times f_A$. On the other hand, we shall show that the totality of such diffeomorphisms includes an open set in $\mathrm{Diff}^1(\mathbf{T}^4)$ the elements of which are not structurally stable (and, consequently, not Ω-stable, since topological transitivity implies that all the points of the manifold are nonwandering). We shall roughly indicate the argument. We remark that this argument is very close to that used by Smale in proving the absence of structural stability in his own first example [26] (see Example 14 in Kušnirenko's lectures), and is in some respects even simpler than Smale's argument.

Let φ_0 be a fixed point of f_A^2, distinct from 0.

PROPOSITION 7.4. *Suppose that the hypotheses of Proposition 7.3 are satisfied, and, in addition, that the diffeomorphism f^{φ_0} in (7.1) has an attractive fixed point ψ_0. Then any diffeomorphism f' which is sufficiently close to f in the C^1-topology is not structurally stable.*

The point (φ_0, ψ_0) is a hyperbolic fixed point of f, with a one-dimensional unstable manifold. The diffeomorphism f' also has a hyperbolic fixed point (φ_0', ψ_0'), close to (φ_0, ψ_0), with a one-dimensional unstable manifold $W_{f'}^u(\varphi_0', \psi_0')$. Furthermore, the torus $T = g_{f'}(\{0\} \times \mathbf{T}^2)$ is invariant with respect to f', the restriction of f' to T is an Anosov diffeomorphism close to f_A, and T is a hyperbolic set of f'. The stable manifold of each point of T is two-dimensional, and the union of these manifolds forms a smooth three-dimensional manifold $W_{f'}^s(T)$. The two manifolds $W_{f'}^u(\varphi_0', \psi_0')$ (one-dimensional) and $W_{f'}^s(T)$ (three-dimensional) intersect transversally, but the points of transversal intersection vary continuously with f' and, in particular, they may belong to the stable manifold of some periodic point of T, and may also belong to the stable manifold of a nonperiodic point. Clearly, the situation described here has a topologically invariant meaning, which shows the absence of structural stability. The diligent reader can easily supply a rigorous argument by analogy with the arguments used in Example 14 of Kušnirenko's lectures.

Appendix. Periodic points and topological entropy

Let X be a compact metric space and $g\colon X \longrightarrow X$ a homeomorphism. The topological entropy $h_{\text{top}}(g)$ of g is defined in the following manner.

Let $\mathfrak{A} = (A_1, \ldots, A_k)$ be a finite open covering of X, let $R(\mathfrak{A})$ be the minimal number of elements in a subset of $\mathfrak{A}$ which still covers the space X, and let $H(\mathfrak{A}) = \log R(\mathfrak{A})$ and $g\mathfrak{A} = \{g\mathfrak{A}_1, \ldots, g\mathfrak{A}_m\}$. If $\mathfrak{A}_1$ and $\mathfrak{A}_2$ are two finite coverings, let $\mathfrak{A}_1 \cdot \mathfrak{A}_2$ denote the covering whose elements are all possible nonempty intersections of sets $A \in \mathfrak{A}_1$ and $B \in \mathfrak{A}_2$. Similarly, one defines the covering $\mathfrak{A}_1 \cdot \mathfrak{A}_2 \cdots \mathfrak{A}_n$. We write

$$h(\mathfrak{A}, g) = \overline{\lim_{n\to\infty}} \frac{1}{n} H(\mathfrak{A} \cdot g\mathfrak{A} \cdots g^{n-1}\mathfrak{A})$$

and, finally, we define $h_{\text{top}}(g)$ as the least upper bound of the $h(\mathfrak{A}, g)$, taken over all possible finite open coverings $\mathfrak{A}$.

The proof of the following theorem constitutes the purpose of this Appendix.

THEOREM 1. *Let Λ be a locally maximal hyperbolic set of the diffeomorphism $f\colon U \longrightarrow M$, and let N_n be the number of fixed points of f^n which belong to Λ. Then*

$$\overline{\lim_{n\to\infty}} \frac{1}{n} \log N_n = h_{\text{top}}(f|_\Lambda).$$

If, in addition, $f|_\Lambda$ has the domain mixing property, then

$$\lim_{n\to\infty} \frac{1}{n} \log N_n = h_{\text{top}}(f|_\Lambda).$$

This theorem follows from results obtained by Bowen [24], [25], using, in particular, the Markov partitions introduced by Sinaĭ [15]. A similar assertion for Anosov flows was proved somewhat earlier by Margulis (see [34], where more delicate results are obtained, giving not only the first but also the two following asymptotic terms of the number of periodic trajectories). We shall present Anosov's proof, which is based upon the appropriate use of the theorem on families of ϵ-trajectories and is considerably simpler than other proofs.

We shall require an alternative definition of topological entropy. Let ρ be a metric in the space X, and let $K_\epsilon(X, \rho)$ denote the greatest number of points of X which are at distances greater than ϵ from each other. Let $g\colon X \longrightarrow X$ be a homeomorphism. We introduce the series of metrics

$$\rho_n(x, y) = \max_{0 \leqslant i \leqslant n} \rho(g^i x, g^i y)$$

in X.

THEOREM 2 (DINABURG [35], §1).

$$h_{\mathrm{top}}(g) = \lim_{\epsilon \to 0} \lim_{n \to \infty} \frac{1}{n} \log K_\epsilon(X, \rho_n).$$

LEMMA. *Let $\delta_1 > 0$ be a number such that, for any two distinct points $x, y \in X$, there is an integer n for which $\rho(g^n x, g^n y) > \delta_1$. Then*

$$\lim_{\epsilon \to 0} \lim_{n \to \infty} \frac{1}{n} \log K_\epsilon(X, \rho_n) = \lim_{n \to \infty} \frac{1}{n} \log K_{\delta_1}(X, \rho_n).$$

PROOF OF THE LEMMA. In the first place, we observe that

$$d(\epsilon) = \lim_{n \to \infty} \frac{1}{n} \log K_\epsilon(X, \rho_n)$$

is a monotonic decreasing function of ϵ. Therefore, it suffices to prove that $d(\epsilon) \leqslant d(\delta_1)$. By the compactness of the space X, for any ϵ there is an $N(\epsilon)$ such that $\rho(x, y) > \epsilon$ implies that $\rho(f^m x, f^m y) > \delta_1$ for some m with $|m| < N(\epsilon)$. Hence

$$K_\epsilon(X, \rho_n) \leqslant K_{\delta_1}(X, \rho_{n+N(\epsilon)})$$

for $n > N(\epsilon)$. From this follows the assertion of the lemma, since $N(\epsilon)$ is independent of n.

We now proceed to prove Theorem 1.

1. We first show that $\overline{\lim}\, n^{-1} \log N_n \leqslant d(\delta_1)$, where δ_1 satisfies the conditions of the lemma (such a δ_1 exists, by virtue of Theorem 2.4). In fact, let $x, y \in \Lambda$ be distinct fixed points of f^n. Since $\rho(f^i x, f^i y) > \delta_1$ for some i, $0 \leqslant i < n$, we have $\rho(x, y) > \delta_1$. Therefore $K_{\delta_1}(\Lambda, \rho_n) \geqslant N_n$.

2. Using Theorem 5.3, we represent Λ in the form $\Lambda = \bigcup_1^m \Lambda_i$. Then $h_{\mathrm{top}}(f|_\Lambda) = \max_i h_{\mathrm{top}}(f|_{\Lambda_i})$ (see [36]); hence, to complete the proof of the theorem, it suffices to prove the inequalities

$$\lim \frac{1}{n} \log N_n^i \geqslant \lim K_{\delta_1}(\Lambda_i, \rho_n), \qquad i = 1, \ldots, m,$$

where N_n^i is the number of fixed points of f^n belonging to Λ_i. Since the action of f on Λ_i is topologically transitive, given any $\epsilon > 0$ there is an $N(\epsilon)$ such that, for any two balls U_1 and U_2 of radius ϵ, $(f^n U_1) \cap U_2 \neq \varnothing$ for some $n < N(\epsilon)$.

In the theorem on families of ϵ-trajectories, take $\delta = \delta_1/4$, and let $\epsilon = \epsilon(\delta_1/4)$ be the corresponding number satisfying the assertion of that theorem. In the space Λ_i with the metric ρ_n, we consider the system of $K = K_{\delta_1}(\Lambda_i, \rho_n)$ points $x_1, \dots, x_K$, at distances greater than δ_1 from each other. We now construct points $x_1', \dots, x_K' \in \Lambda_i$ such that $\rho(x_j', f^n x_j) < \epsilon$ and $\rho(f^{n_j} x_j', x_j) < \epsilon$, where $n_j < N(\epsilon) = N$, $j = 1, \dots, K$, and for each j we consider the sequence of points

$$x_j, fx_j, \dots, f^n x_j, fx_j', \dots, f^{n_j - 1} x_j'.$$

Let $X^j = \{0, 1, \dots, n + n_j - 1\}$,

$$g^j\colon X^j \longrightarrow X^j,\ g(l) = l + 1 \pmod{n + n_j}, \quad \varphi_j\colon X^j \longrightarrow M,$$

$$\varphi_j(l) = \begin{cases} f^l x_j, & 0 \leqslant l \leqslant n, \\ f^{l-n} x_j', & n + 1 \leqslant l \leqslant n + n_j - 1. \end{cases}$$

Applying the theorem on families of ϵ-trajectories to this situation, we obtain points x_j'' such that

$$\rho(f^l x_j'', \varphi_j(l)) < \delta_1/4, \qquad f^{n+n_j} x_j'' = x_j''.$$

By the local maximality of the set Λ_i, $x_j'' \in \Lambda_i$. Clearly, the points x_j'', $j = 1, \dots, K$, are pairwise distinct. Therefore $\Sigma_{l=0}^{N-1} N_{n+l} \geqslant K_{\delta_1}(\Lambda_i, \rho_n)$, and consequently there is an $l_n < N$ such that

$$N_{n+l_n} \geqslant \frac{K_{\delta_1}(\Lambda_i, \rho_n)}{N};$$

that is,

$$\frac{1}{n + l_n} \log N_{n+l_n} = \frac{n}{n + l_n} \left(\frac{1}{n} \log K_{\delta_1}(\Lambda_i, \rho_n) \right) - \frac{1}{n + l_n} \log N.$$

Since N is independent of n,

$$\varlimsup_{n \to \infty} \frac{1}{n} \log N_n \geqslant \lim_{n \to \infty} \frac{1}{n} \log K_{\delta_1}(\Lambda_i, \rho_n).$$

Thus the first part of Theorem 1 is proved. If also $f|_\Lambda$ has the mixing property, one may take $n_j = N$ in the preceding argument; hence $N_{n+N} \geqslant K_{\delta_1}(\Lambda, \rho_n)$ and

$$\frac{1}{n} \log N_n \geqslant \frac{1}{n} \log K_{\delta_1}(\Lambda, \rho_{n-N}) = \frac{n - N}{n} \left(\frac{1}{n - N} \log K_{\delta_1}(\Lambda, \rho_{n-N}) \right).$$

BIBLIOGRAPHY

1. D. V. Anosov, *Geodesic flows on closed Riemannian manifolds with negative curvature*, Trudy Mat. Inst. Steklov. 90 (1967); English transl. in Proc. Steklov Inst. Math. 90 (1967)(1969).

2. S. Smale, *Differentiable dynamical systems*, Bull. Amer. Math. Soc. 73 (1967), 747–817; Russian transl., Uspehi Mat. Nauk 25 (1970), no. 1 (151), 113–185.

3. John Franks, *Anosov diffeomorphisms*, Global Analysis, Proc. Sympos. Pure Math., vol. 14, Amer. Math. Soc., Providence, R.I., 1970, pp. 61–93.

4. Stephen Smale, *Diffeomorphisms with many periodic points*, Differential and Combinatorial Topology (A Symposium in Honor of Marston Morse), Princeton Univ. Press, Princeton, N.J., 1965, pp. 63–80.

5. John N. Mather, *Characterization of Anosov diffeomorphisms*, Nederl. Akad. Wetensch. Proc. Ser. A 71 = Indag. Math. 30 (1968), 479–483.

6. J. W. Robbin, *On structural stability*, Bull. Amer. Math. Soc. 76 (1970), 723–726.

7. John Franks, *Differentiably Ω-stable diffeomorphisms*, Topology 11 (1972), 107–113.

8. Nelson Dunford and Jacob T. Schwartz, *Linear operators*. Vol. I, Interscience, New York, 1958.

9. D. V. Anosov, *A certain class of invariant sets of smooth dynamical systems*, Proc. Fifth Internat. Conf. Nonlinear Oscillations (Kiev, 1969), Vol. 2, Izdanie Inst. Mat. Akad. Nauk Ukrain. SSR, Kiev, 1970, pp. 39–45. (Russian)

10. Serge Lang, *Introduction to differentiable manifolds*, Interscience, New York, 1962; Russian transl., "Mir," Moscow, 1967.

11. J. Dieudonné, *Foundations of modern analysis*, Academic Press, New York, 1960.

12. V. M. Alekseev, *Quasirandom dynamical systems*, Doctoral dissertation, Moscow State Univ., 1969; abstract published in Mat. Zametki 6 (1969), 489–498; English transl. in Math. Notes 6 (1969).

13. Morris W. Hirsch and Charles C. Pugh, *Stable manifolds and hyperbolic sets*, Global Analysis, Proc. Sympos. Pure Math., vol. 14, Amer. Math. Soc., Providence, R.I., 1970, pp. 133–163.

14. Zbigniew Nitecki, *Differentiable dynamics*, M.I.T. Press, Cambridge, Mass., 1971.

15. Ja. G. Sinaĭ, *Markov partitions and Y-diffeomorphisms*, Funkcional. Anal. i Priložen. 2 (1968), no. 1, 64–89; English transl. in Functional Anal. Appl. 2 (1968).

16. G. A. Margulis, *Certain measures that are connected with Y-flows on compact manifolds*, Funkcional. Anal. i Priložen. 4 (1970), no. 1, 62–76; English transl. in Functional Anal. Appl. 4 (1970).

17. Ja. G. Sinaĭ, *Dynamical systems with countably-multiple Lebesgue spectrum*. II, Izv. Akad. Nauk SSSR Ser. Mat. 30 (1966), 15–68; English transl. in Amer. Math. Soc. Transl. (2) 68 (1968).

18. ———, *Mesures invariantes des Y-systèmes*, Proc. Internat. Congr. Math. (Nice, 1970), vol. 2, Gauthier-Villars, Paris, 1971, pp. 929–940.

19. S. E. Newhouse, *On codimension one Anosov diffeomorphisms*, Amer. J. Math. 92 (1970), 761–770.

20. John Franks, *Anosov diffeomorphisms on tori*, Trans. Amer. Math. Soc. 145 (1969), 117–124.

21. V. M. Alekseev, *Perron sets and topological Markov chains*, Uspehi Mat. Nauk 24 (1969), no. 5 (149), 227–228. (Russian)

22. S. Smale, *The Ω-stability theorem*, Global Analysis, Proc. Sympos. Pure Math., vol. 14, Amer. Math. Soc., Providence, R.I., 1970, pp. 289–297.

23. M. Hirsch et al., *Neighborhoods of hyperbolic sets*, Invent. Math. 9 (1969/70), 121–134.

24. Rufus Bowen, *Periodic points and measures for Axiom A diffeomorphisms*, Trans. Amer. Math. Soc. 154 (1971), 377–397.

25. ———, *Markov partitions for Axiom A diffeomorphisms*, Amer. J. Math. 92 (1970), 725–747.

26. S. Smale, *Structurally stable systems are not dense*, Amer. J. Math. 88 (1966), 491–496.

27. R. Abraham and S. Smale, *Nongenericity of Ω-stability*, Global Analysis, Proc. Sympos. Pure Math., vol. 14, Amer. Math. Soc., Providence, R.I., 1970, pp. 5–8.

28. M. W. Hirsch, C. C. Pugh and M. Shub, *Invariant manifolds*, Bull. Amer. Math. Soc. 76 (1970), 1015–1019.

29. Carl P. Simon, *On a classification of a Baire set of diffeomorphisms*, Bull. Amer. Math. Soc. **77** (1971), 783–787.

30. Sheldon E. Newhouse, *Nondensity of Axiom* A(a) *on* S^2,* Global Analysis, Proc. Sympos. Pure Math., vol. 14, Amer. Math. Soc., Providence, R.I., 1970, pp. 191–202.

31. Charles C. Pugh, *The closing lemma*, Amer. J. Math. **89** (1967), 956–1009.

32. S. Smale, *Global stability questions in dynamical systems*, Lectures in Modern Analysis and Applications. I, Springer-Verlag, Berlin and New York, 1970, pp. 150–158.

33. J. Palis and S. Smale, *Structural stability theorems*, Global Analysis, Proc. Sympos. Pure Math,, vol. 14, Amer. Math. Soc., Providence, R.I., 1970, pp. 223–231.

34. G. A. Margulis, *On some questions in the theory of* $\mathcal{Y}$*-systems,* Dissertation, Moscow State Univ., Moscow, 1970. (Russian)

35. E. I. Dinaburg, *On the topological entropy of dynamical systems*, Dissertation, Moscow, State Univ., Moscow, 1971. (Russian)

36. R. L. Adler, A. G. Konheim and M. H. McAndrew, *Topological entropy*, Trans. Amer. Math. Soc. **114** (1965), 309–319.

37. M. I. Brin and Ja. B. Pesin, *Partially hyperbolic dynamical systems*, Izv. Akad. Nauk SSSR Ser. Mat. **38** (1974), 170–212; English transl. in Math. USSR Izv. **8** (1974).

38. M. I. Brin, *Topological transitivity of a certain class of dynamical systems, and flows of frames on manifolds of negative curvature*, Funkcional. Anal. i Priložen. **9** (1975), no. 1, 9–19; English transl. in Functional Anal. Appl. 9 (1975).

39. Richard L. Bishop and Richard J. Crittenden, *Geometry of manifolds*, Academic Press, New York, 1964.

40. Shlomo Sternberg, *Lectures on differential geometry*, Prentice-Hall, Englewood Cliffs, N.J., 1964.

41. V. V. Nemyckiĭ and V. V. Stepanov, *Qualitative theory of differential equations*, 2nd ed., GITTL, Moscow, 1949; English transl., Princeton Univ. Press, Princeton, N.J., 1960.

Translated by E. J. BRODY

**Editor's note.* The title given in the original is *On generic properties of differentiable automorphisms,* which is Newhouse's unpublished thesis.

Amer. Math. Soc. Transl.
(2) Vol. 116, 1981

QUASIRANDOM OSCILLATIONS AND QUALITATIVE QUESTIONS IN CELESTIAL MECHANICS*

V. M. ALEKSEEV

Contents

INTRODUCTION

Among the various areas of science where mathematics finds its applications and is a valued guest, celestial mechanics occupies a quite exceptional position. For a long time, almost 300 years, its problems, and especially the "three-body problem", have served as a proving ground on which generation after generation of mathematicians has tested the power of techniques of investigation, both old ones and those newly devised. The naturalness of the problems, and the simplicity of their formulation, is combined with the unusual richness and variety of their content. Attempts to solve the three-body problem, or to obtain at least partial results on the properties of its solutions, have been the starting point for the development of many areas of science. Many results have even now not yet lost their importance, while others have only historical interest; the number of publications is measured in the hundreds, if not in the thousands.

At the turn of the century Henri Poincaré summed up in his "Méthodes nouvelles de la mécanique céleste" the results of investigations during the 19th century, and gave abundant stimulus to scientists of the 20th. One is amazed at the wealth of ideas and concepts contained in this book: the small parameter method, asymptotic

* Translation of Ninth Math. Summer School (Kaciveli, 1971), Izdanie Inst. Mat. Akad. Nauk Ukrain. SSR, Kiev, 1972, pp. 212–341. MR 53 #11667.
1980 *Mathematics Subject Classification.* Primary 34C15, 34F05, 70F15; Secondary 70F07.

series, the investigation of periodic solutions, homoclinic curves, integral invariants and so on. In a word, a considerable part of what is now called the "qualitative theory of differential equations" arises from Poincaré's "Méthodes nouvelles" and consequently from celestial mechanics.

The adoption of qualitative methods in the theory of nonlinear oscillations and in other problems of nonlinear mechanics took place later, but it happened to be primarily these problems that began to be connected with the qualitative theory. Many of these problems lead to the investigation of flows on two-dimensional manifolds, where the situation is appreciably simpler (see, for example, the lectures of A. G. Kušnirenko contained in this volume), but in return the progress has been more substantial and the methods more fruitful.

The two-dimensional qualitative theory of differential equations has received considerable development, though it is still clearly incomplete, as is constantly demonstrated in conferences and symposia. Presently, the qualitative theory is developing quite vigorously; and, in my opinion, every return to its sources, among which I count celestial mechanics, will give new strength and freshness.

One such return took place when A. N. Kolmogorov, and then V. I. Arnol'd and Jürgen Moser, obtained their remarkable results relating to invariant tori and conditional periodic motions in Hamiltonian systems. One need hardly recall how important this progress has been in what Poincaré termed the "fundamental problem of dynamics" and how stimulating its influence was on qualitative theory.

Now, it seems to me, the ground is being prepared for yet another such return, one which may already be under way. We hope that this return will prove sufficiently fruitful.

In addition to materials from the Ukrainian Mathematics Summer School in Kaciveli (June, 1971), I have used the Russian version of the report [35], and also results I became aware of relatively recently. Part I contains a survey of certain qualitative investigations in problems of celestial mechanics. Part II is a descriptive introduction to "symbolic dynamics"; I have attempted to keep technical details from obscuring the essence of the matter. In Part III I wanted to show how the methods of symbolic dynamics can prove useful in even the simplest problems of the theory of nonlinear oscillations, and what they permit us to do in application to the problems considered in Part I. I convey my deep gratitude to the organizers of the School, who invited me to give these lectures, and who showed uncommon patience in awaiting the written version.

PART I

ON CERTAIN MATHEMATICAL PROBLEMS IN CELESTIAL MECHANICS

§1. The basic problem of celestial mechanics

It is well known that classical celestial mechanics is concerned mainly with various aspects of the so-called "n-body problem". This problem consists in studying the motion of n material points P_i attracting one another according to Newton's law.([1]) Denoting by m_i the mass of the point P_i, by r_i its radius-vector and by γ the gravitational constant, we have the following system of equations of motion:

$$\ddot{r}_i = \sum \frac{\gamma m_j r_{ji}}{|r_{ji}|^3}; \tag{1}$$

for brevity, we write $r_{ji} = r_j - r_i$. These equations determine a dynamical system (a phase-flow) in a $6n$-dimensional phase space, to be obtained from $\mathbf{R}^{6n}$ by removing $\frac{1}{2}n(n-1)$ planes $r_{ji} = 0$ of codimension 3.

The center of mass of the system in question moves uniformly in a straight line, so that, if we choose it as the location of an inertial system of coordinates, we shall have identically

$$\sum_i m_i r_i \equiv 0. \tag{2}$$

These relations make it possible to reduce the dimensionality of the phase space to $6n - 6$ by eliminating, for example, the coordinates and velocity of one of the points P_i. In addition, the system (1) has 4 further algebraic integrals: the energy integral

$$\frac{1}{2} \sum_i m_i |\dot{r}_i|^2 - \sum_{i<j} \frac{\gamma m_i m_j}{|r_{ij}|} \equiv h, \tag{3}$$

and three area integrals (components of the angular momentum vector)

$$\sum_i m_i r_i \times \dot{r}_i \equiv C. \tag{4}$$

Along with the "general problem", in which all masses are supposed positive, one considers also limiting cases when some of the masses m_i in (1) are set equal to zero. In the language of physics this means that we neglect the influence of the bodies in question on the motion of the remainder. In this situation one usually speaks of the "restricted problem". Particularly well known is the problem of the motion of a body of zero mass (a "planetoid" or an "asteroid") in a gravitational field created by two bodies rotating in circular orbits about their common center of mass, in which all three bodies remain for all time in one and the same plane. Strictly speaking, it was this problem that Poincaré termed the "restricted three-body problem", but now it is often termed more diffusely "the restricted plane circular problem" in distinction to "the restricted elliptic problem", and so forth. If one equates to zero all the masses except one, we get the "ideal planetary system", in which bodies of zero mass (the "planets") rotate around one body (the "sun") along purely Keplerian orbits, not exerting any influence on each other. In the classical perturbation theory of celestial mechanics this case appears as the zero-order approximation.

([1]) The so-called "generalized n-body problem", in which the law of attraction differs from the Newtonian one (of the inverse square), will not be dealt with here. In particular, we shall not touch on the series of works by the late Kiev mathematician Ju. D. Sokolov, who did much in this direction.

Another limit case is obtained when the position of certain bodies is fixed; here one usually speaks of "immovable centers of attraction". In particular, one may consider the simplest–the problem of the motion of one material point in the field of a single attracting center. In the sequel we will term this case the "problem of Kepler"; it scarcely needs mention that the integration was already carried out by Newton, and that the qualitative and quantitative analysis of the solutions lies at the base of all celestial mechanics. Nevertheless, I hope to show in §2 that even here it is possible, if not to obtain new results, at least to see old ones from a new, unexpected point of view.

The problem of the motion of a material point in the field of two attracting centers also belongs among those integrable by quadratures [14]. Quite recently, interest in this case quickened considerably as it turned out that it is a good approximation for the problem of the motion of a satellite in the gravitational field of a planet which is not strictly spherical. If the planet is drawn out like a cucumber this is not surprising, but how if, like the real Earth, it is an oblate spheroid? It turns out that in this case one must locate the fixed centers in complex conjugate points of space, even though the problem is considered in the purely real domain (an exposition of these interesting and beautiful results of E. P. Aksenov, E. A. Grebenikov and V. P. Demin is to be found in [3]).

The two-body problem is easily reduced to the problem of Kepler, and accordingly the equations (1) for $n = 2$ can be completely integrated. On the other hand, for $n \geqslant 3$ they cannot be explicitly integrated (by quadratures).(2) At the same time, the practical requirements of astronomy have long since prompted the development of numerical methods for their solution. The very character of the motion of the heavenly bodies–approximately periodic or sufficiently closely approximable by the superposition of several periodic motions–makes it natural to construct solutions in the form of a multiple Fourier series

$$r(t) = \sum A_m \begin{matrix}\sin\\ \cos\end{matrix}((m_1\omega_1 + \cdots + m_n\omega_n)t). \tag{5}$$

Numerous "theories of motion" (of the moon, the planets, asteroids, satellites of the planets, and so on) lead in the end to the construction of partial sums of series analogous to (5), giving an approximation suitable for comparison with given observations. Many investigations of the 19th century are devoted to the development of various procedures enabling one to avoid the appearance in purely trigonometric expansions (5) of "secular" terms of the form t^m or $t \sin t$. Poincaré gave in [8] a detailed analysis of these procedures; he in fact showed that the series thus obtained were in general divergent, although their partial sums did indeed give an approximation to the true solution over finite time-intervals. At the same time, it became clear from the work of Poincaré that it might not be hopeless to attempt to solve the problem within the framework of the general theory of analytic functions. In fact, in 1912, K. F. Sundman

(2) Perhaps it would be more cautious to say "for the present have not been integrated".

published his remarkable theorem, which was perceived as a final solution of the three-body problem.

We will return to Sundman's theorem again in §3, for the time being merely stating that, while having indisputable theoretical interest, it yields little for practical purposes. Moreover, this theorem essentially deals only with an individual solution of the three-body problem, and does not clarify the global structure of the phase-flow. Hence, from the point of view of contemporary mathematics, the three-body problem remains just as interesting as it was a hundred years ago.

The vigorous development of computational technology and the appearance of new problems in celestial mechanics, together with space navigation and the motion of artificial satellites, necessitated a review of the classical approximative methods (the evolution of orbits must be monitored over thousands of revolutions, one must take into account departures of the gravitational field from a strictly Newtonian field arising from the asymmetry of the Earth, and so on). In many cases direct methods for the approximate integration of the differential equations are suitable. However, computers can also operate successfully in combination with traditional techniques. Recently, extensions of the series method (5) have been obtained, in which the operations are programmed directly in terms of letters, and not with numerical coefficients, the machine thus constructing an "analytic theory".

On the other hand, in the spirit of the 20th century interest in qualitative problems, to which these lectures are basically devoted, one should admit at once that at present we do not have at our disposal a large stock of conclusions regarding the qualitive properties of the solutions of the problem for general n for $n \geqslant 4$; the greater part of them can be seen in [11], [13] and [14]. Of works in recent years we mention [48] and [49]. The remarkable paper of V. I. Arnol'd [18] and the recent paper of Stephen Smale [31] must be specially noted: their significance is by no means confined to celestial mechanics, and the results relating to the n-body problem are but one of many possible applications of the general theory. We shall also consider some of the qualitative results, both those fairly old and also some relatively recent, dealing mainly with the cases $n = 2$ and 3.

§2. Geometrical interpretation of the Kepler problem

If $n = 2$ and the center of mass of the bodies P_1 and P_2 is at rest, in conformity with (2), at the origin, the system (1) can be simplified. In fact, eliminating r_2 with the aid of (2), we obtain for r_1 the same equation as if the body P_1 moved in the Newtonian gravitational field of an immovable mass $\mu = m_2^2/(m_1 + m_2)^2$, located at the origin (the "Kepler problem"). With the natural choice of units $m_1 = 1$ and $\gamma_\mu = 1$, for $r = r_1$ we have

$$\ddot{r} = -r/|r|^3. \tag{6}$$

The subsequent argument does not in fact depend on the number of coordinates, so that in (6) r may be taken to be a vector in $\mathbf{R}^m$. For simplicity, however, we confine ourselves to the case $m = 2$.

The phase space for the Kepler problem has (for $m = 2$) the dimension 4. Fixing the values of the energy integral h and the momentum C, we get a two-dimensional manifold. For $h < 0$ and $C \neq 0$ it will be compact. If we replace (6) by the equation of motion in an arbitrary field of force

$$\ddot{r} = -f(|r|)r, \tag{7}$$

then in the typical case such a manifold will be a two-dimensional torus, and every trajectory on it will form an everywhere dense winding. Projecting the trajectory from phase space into coordinate space, we find that a typical orbit, which does not go off to infinity, of a material point governed by (7) will fill out a certain annulus (Figure 1) in an everywhere dense manner. In exceptional resonance cases the invariant tori will decompose into a family of periodic trajectories.

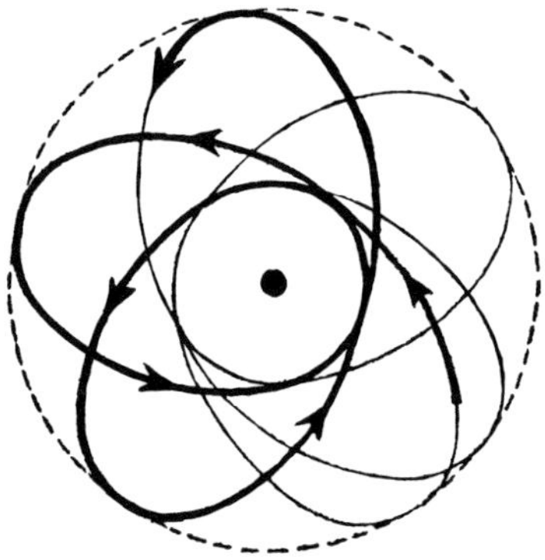

FIGURE 1

If however the attractive force of the immovable center decreases like $1/r^2$, then the exceptional case becomes the rule, and degeneration takes place. An orbit which does not go off to infinity is always an ellipse (Figure 2), and in the phase space all two-dimensional tori decompose into families of periodic trajectories.

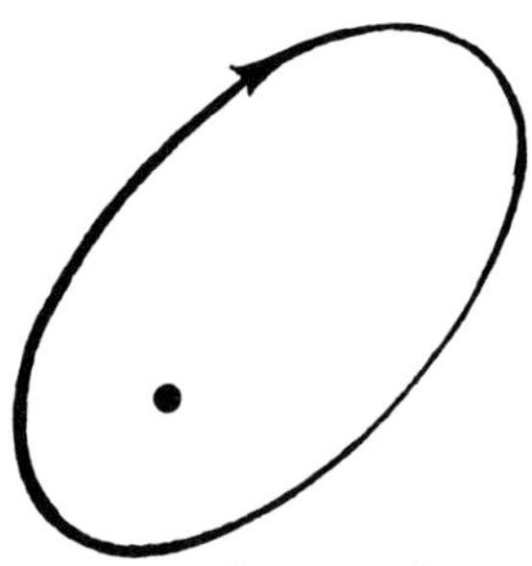

FIGURE 2

It is natural to expect that such degeneration owes its origin to the presence of another first integral ([4], §50). It is known that the classical integrals (3) and (4) are connected (by the theorem of E. Noether) with the invariance of the equations (1) under shifts of time (the energy integral) and under rotations (the angular momentum integral). This suggests the thought that the new integral must also be a consequence of some "hidden symmetry" of the Kepler problem.

The geometrical interpretation of the case $h < 0$ given by Moser in [44] makes it possible to show this symmetry explicitly; but then Moser notes that as far back as 1935 V. Fock [41] used a transformation, crucial for the present considerations, in order to clarify the degeneration of energy-levels in the quantum model of the hydrogen atom (an analogous effect in physics is likewise connected with "hidden symmetry"). Ju. S. Osipov [25] extended Moser's considerations to the case of arbitrary n, and gave a somewhat different proof, which with minor variations will be carried out in the sequel.

We recall first the definition of geodesic flow on a Riemannian manifold M^n (see, for example, [17]). By this one means a dynamical system T_s acting in a manifold W^{2n-1} of unit tangent vectors to M^n.

Let τ be a unit tangent vector at the point $p \in M^n$. We draw through p an (oriented) geodesic in the direction of τ and, marking off on it an arc length s, obtain the point p' and the unit vector τ', touching there the same (oriented) geodesic. By definition, the mapping T_s takes $\tau \in W^{2n-1}$ into $\tau' = T_s\tau$ (Figure 3).

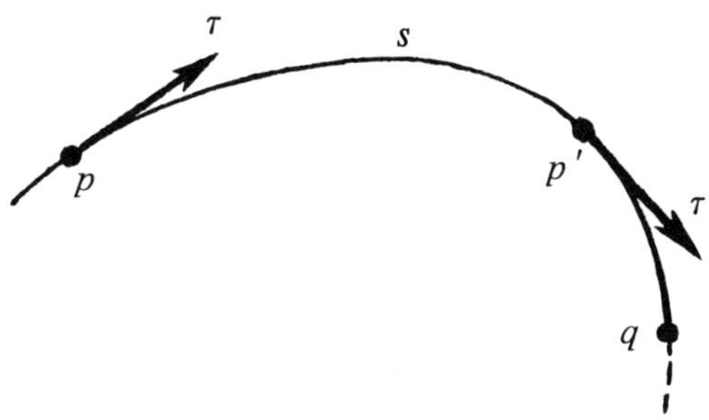

FIGURE 3

THEOREM. *The phase flow of the plane Kepler problem* (6) *on the manifold of constant energy h is, to within a time-change, equivalent to geodesic flow on a surface of constant Gaussian curvature* $-2h$. *After the addition of a single point, this surface is isometric to a sphere if* $h < 0$, *the euclidean plane if* $h = 0$, *and the Lobachevsky plane if* $h > 0$.

As already stated, an analogous assertion holds for any dimension m. We recommend the reader to carry out the appropriate modifications in the statement.(3)

PROOF. We make a time-change, introducing a new independent variable s by the relation

$$t = \int |r|\, ds. \tag{8}$$

(3) Moser [44] proved this assertion for $h < 0$. When I mentioned this result in Kaciveli, A. D. Myškis conjectured the validity of an analogous result for all h, which was in face later proved in [25].

In the canonical variables $p = (p_1, p_2) = dr/dt$ and $r = (r_1, r_2)$ the equations of the Kepler problem have the Hamiltonian form

$$\frac{dr}{dt} = \frac{\partial H}{\partial p}, \qquad \frac{dp}{dt} = -\frac{\partial H}{\partial r}, \tag{9}$$

where $H(p, r) = p^2/2 - 1/|r|$. On the constant-energy integral manifold $H(p, r) \equiv h$ one may transform the equations (9) as follows:

$$\frac{dr}{ds} = |r|\frac{dr}{dt} = |r|\frac{\partial H}{\partial p} = \frac{\partial}{\partial p}[|r|(H-h)],$$

$$\frac{dp}{ds} = |r|\frac{dp}{dt} = -|r|\frac{\partial H}{\partial r} = -\frac{\partial}{\partial r}[|r|(H-h)]$$
$$+\frac{\partial |r|}{\partial r}(H-h) = -\frac{\partial}{\partial r}[|r|(H-h)],$$

since $H - h$ vanishes identically on this manifold. On the same manifold

$$0 = |r|(H-h) = |r|\left(\frac{p^2}{2} - \frac{1}{|r|} - h\right) = \frac{|r|(p^2-2h)}{2} - 1,$$

which renders possible the further transformation of the equations:

$$\frac{dr}{ds} = \frac{\partial}{\partial p}\,|\,|r|(H-h)| = \frac{|r|(p^2-2h)}{2}\frac{\partial}{\partial p}|\,|r|(H-h)|$$
$$= \frac{|r|(p^2-2h)}{2}\frac{\partial}{\partial p}\left(\frac{|r|(p^2-2h)}{2}\right)$$
$$= \frac{\partial}{\partial p}\left|\frac{(p^2-2h)^2 r^2}{8}\right|$$

and analogously

$$\frac{dp}{ds} = \frac{\partial}{\partial r}\left[\frac{(p^2-2h)^2 r^2}{8}\right].$$

Thus the change of time (8) transforms the Hamiltonian equations (9) on the constant-energy surface $H = h$ into the Hamiltonian equations

$$\frac{dr}{ds} = \frac{\partial F}{\partial p}, \qquad \frac{dp}{ds} = -\frac{\partial F}{\partial r},$$

$$F(p, r) = \frac{(p^2-2h)^2 r^2}{8} \tag{10}$$

on the surface $F \equiv ½$. We remark that for $h > 0$ the surface $F = ½$ consists of two "poles"; according to the above transformations, one must choose the one for which $p^2 > 2h$.

We now eliminate r from (10):

$$\frac{dp}{ds} = -\frac{\partial F}{\partial r} = -\frac{(p^2 - 2h)^2 r}{4};$$

$$r = -\frac{4}{(p^2 - 2h)^2}\frac{dp}{ds};$$

$$-\frac{d}{ds}\left[\frac{4}{(p^2 - 2h)^2}\frac{dp}{ds}\right] = \frac{dr}{ds} = \frac{\partial F}{\partial p}$$

$$= \frac{(p^2 - 2h)r^2 p}{2} = \frac{8p}{(p^2 - 2h)^3}\left(\frac{dp}{ds}\right)^2. \tag{11}$$

Hence

$$\frac{d^2 p}{ds^2} = \frac{2}{p^2 - 2h}\left[2\left(p, \frac{dp}{ds}\right)\frac{dp}{ds} - p\left(\frac{dp}{ds}\right)^2\right], \tag{12}$$

where

$$\left|\frac{dp}{ds}\right| = \frac{(p^2 - 2h)^2 |r|}{4} = \frac{p^2 - 2h}{2}.$$

The manifold $F = ½$ may be projected onto the whole (p_1, p_2)-space in the case $h < 0$, onto the region $p \neq 0$ in the case $h = 0$ (taking account of the choice of pole), and onto the region $p^2 > 2h$ in the case $h > 0$. These open subsets in $\mathbf{R}^2$ will for brevity be denoted by M_h. We introduce in M_h a new Riemannian metric so that the velocity-vector dp/ds will be a unit vector:

$$d\sigma^2 = \frac{4dp^2}{(p^2 - 2h)^2}. \tag{13}$$

It is not hard to see that (12) constitute the equations of geodesic lines of the new metric, and, by the construction itself, the parameter s has the meaning of arc-length.

By (13), the components of the co- and contravariant metric tensors (10) are respectively equal to

$$g_{ij} = \frac{4\delta_{ij}}{(p^2 - 2h)^2}; \qquad g^{ij} = \frac{(p^2 - 2h)^2}{4}\delta_{ij};$$

whence we find the Christoffel symbols:

$$\Gamma^l_{km} = \frac{1}{2}\sum_s g^{sl}\left[\frac{\partial g_{sm}}{\partial p_k} + \frac{\partial g_{sk}}{\partial p_m} - \frac{\partial g_{km}}{\partial p_s}\right]$$

$$= -\frac{2}{p^2 - 2h}[\delta_{lm}p_k + \delta_{lk}p_m - \delta_{km}p_l].$$

Consequently, the equations of the geodesics have the form

$$\frac{d^2 p_l}{ds^2} = -\sum_{k,m} \Gamma^l_{km} \frac{dp_k}{ds}\frac{dp_m}{ds}$$

$$= \frac{2}{p^2 - 2h} \sum_{k,m} (\delta_{lm} p_k + \delta_{lk} p_m - \delta_{km} p_l) \frac{dp_k}{ds}\frac{dp_m}{ds}$$

$$= \frac{2}{p^2 - 2h} \left[2\left(p, \frac{dp}{ds}\right)\frac{dp_l}{ds} - p_l\left(\frac{dp}{ds}\right)^2 \right],$$

which coincides with (12).

We consider the mapping φ of the manifold $F = ½$ into the tangent bundle TM_h, given by the equation $\varphi(p, r) = (p, \tau)$, where

$$\tau = \frac{dp}{ds} = -\frac{r}{|r|^2} = -\frac{(p^2 - 2h)^2}{4} r.$$

It is easy to see that this mapping establishes a one-to-one correspondence between the manifold $F = ½$ (or its pole, if $h > 0$) and the manifold W of unit tangent vectors. It remains to check that under the mapping φ the phase flow determined by (10) goes into the geodesic flow T_s.

In fact, the trajectory $(p(s), r(s))$ of the flow (10) transforms into $\varphi(p(s), r(s)) = (p(s), \tau(s))$, where, by what has been proved, $p(s)$ is a geodesic, and $\tau(s) = dp/ds$ is (in the metric (13)) a unit tangent vector to it. The geodesic through the point $p = p(0)$ in the direction $\tau = \tau(0)$ coincides with $p(s)$, and since the vector dp/ds is unit, s is the arc-length along this geodesic. Hence the mapping T_s of the geodesic flow takes $(p(0), \tau(0))$ into $(p(s), \tau(s)) = \varphi(p(s), r(s))$, as asserted.

The metric (13) is well known.

For $h < 0$ the stereographic projection

$$p_i = \frac{x_i \cdot 2|h|}{2 - x_0\sqrt{2|h|}} \tag{14}$$

takes M_h into the punctured sphere $x_0^2 + x_1^2 + x_2^2 = 2x_0(2|h|)^{-½}$ of radius $R = (2 - |h|)^{-½}$, and the metric (13) goes into the usual euclidean one. Thus, the Riemannian manifold M_h is of constant positive (Gaussian) curvature $2|h|$.

For $h = 0$ the inversion

$$p_i = \frac{2x_i}{x_1^2 + x_2^2} \tag{15}$$

takes M_0 again into itself, and (13) becomes the euclidean metric

$$\frac{4(dp_1^2 + dp_2^2)}{(p_1^2 + p_2^2)^2} = dx_1^2 + dx_2^2.$$

For $h > 0$ the inversion

$$p_i = \frac{\sqrt{2h}\, x_i}{x_1^2 + x_2^2} \tag{16}$$

maps M_h onto the interior of the unit disc with punctured center, and the metric (13) goes (to within a constant multiple) into the usual metric of the Poincaré model of the Lobachevsky plane:

$$\frac{4(dp_1^2 + dp_2^2)}{(p_1^2 + p_2^2 - 2h)^2} = \frac{1}{2h} \frac{4(dx_1^2 + dx_2^2)}{(1 - x_1^2 - x_2^2)^2}.$$

This concludes the proof of the theorem.

The equivalence we have proved (to within a time-change (8)) between the phase flow of the problem (6) on the constant-energy manifold $H = h$ and the geodesic flow on the manifold M_h with metric (13) makes it possible, first of all, to carry out a "regularization" of the Kepler problem.

The equation (6) has a singularity at $r = 0$. If the area constant (integral of momentum) $C \neq 0$, then, as is easy to verify, $|r|$ is bounded away from zero by a positive constant, so that the solution does not have singularities and is an analytic function of t on the whole axis $-\infty < t < +\infty$. On the other hand, if $C = 0$, the motion takes place along a line passing through the origin, and necessarily $\lim_{t \to t_0} r(t) = 0$ for some t_0. Without loss of generality we may assume that $t_0 > 0$. In the case of an actual mechanical problem the moving point collides with the fixed center of attaction (in the case of a two-body problem this means the collision of the moving points at their center of mass), and it is senseless to speak of what will happen for $t > t_0$, since for close proximity and all the more at a collision the idealized representation of a physical object by a mathematical point becomes unsuitable. However, from the mathematical point of view it is natural to try to continue $r(t)$ for $t > t_0$ in such a way that the resulting vector-valued function should be in some sense "regular". When speaking here and in the next section of the "continuation of a solution past the instant of collision" we shall always have in view just this purely mathematical formulation of the question.

One may attempt to carry out the regularization of the solution $r(t)$ with the aid of a limit passage, taking a family of solutions $r(t, d)$ of the problem (6), so that for $d \neq 0$ they do not have singularities (i.e. they should be in the region $C \neq 0$) and such that

$$r(t) = \lim_{d \to 0} r(t, d) \tag{17}$$

for $0 \leqslant t < t_0$. If here the right-hand side of (17) exists also for $t > t_0$, then this limit may be considered as a continuation of $r(t)$ past the instant of collision. Another possibility is suggested by the analytic character of the solution. In fact, in the complex plane one may avoid the singular point $t = t_0$ and obtain the values $r(t)$, $t > t_0$, by analytic continuation. Naturally, the question of the physical meaning of "complex time" will be empty: this will be just as much a mathematical abstraction as the very idea of continuation of a motion past the instant of collision.

Since at the singular point $\lim_{t\to t_0} r(t) = 0$, we find from the energy integral that

$$|p(t)| = \sqrt{2h + 2|r(t)|^{-1}} \longrightarrow \infty .$$

Hence to a solution with singularity there corresponds a geodesic in the metric (13) which goes to the point at infinity p_∞ (in the sense of the coordinates (p_1, p_2), but not in the sense of arc-length).

The manifold M_h with the metric (13) is not a complete metric space, but becomes one after completion with the point p_∞. According to (14)–(16), the completed space $\widetilde{M}_h = M_h \cup \{p_\infty\}$ is isometric to a sphere if $h < 0$, to the euclidean plane if $h = 0$ and to the Lobachevsky plane if $h > 0$. The point p_∞ here corresponds to the "north pole" $((2/|h|)^{1/2}, 0, 0)$ of the sphere for $h < 0$, and to the point $x_1 = x_2 = 0$ in the other cases. In all three cases a geodesic arc, ending in some point q, can be uniquely continued to a complete geodesic passing through q (Figure 3). Hence a geodesic of the metric (13) corresponding to the solution of the Kepler problem with singularity has a unique continuation past p_∞. It is easy to see that this continuation constitutes a regularization in the sense of (17).

We now consider the situation from the point of view of analytic function theory. The equations of geodesics on a sphere, the euclidean plane and the Lobachevsky plane do not have real singularities. Hence along a geodesic the coordinates x_i connected with $p = (p_1, p_2)$ by (14)–(16) are analytic (and indeed elementary) functions of the arc-length s, without singular points on the real axis $-\infty < s < +\infty$. "Puncturing" the surface at the image point p_∞ generates a "removable singular point" in the functions $x_i(s)$ (in the standard terminology of the theory of functions of a complex variable). It is clear that avoiding it by a path close to the real axis will lead to the same result as the continuation of a geodesic in the whole manifold $\widetilde{M}_h$. As to the variables $r(t)$ and $p = dr/dt$, there will be a break in the analyticity of their dependence on the "physical time" t, connected with s by the relation (8).

In fact, for $s = 0$ let a geodesic in $\widetilde{M}_h$ pass through the "puncture". Noting that s is the arc-length, and denoting by τ the unit vector tangent to the geodesic at the point of puncture, we have

$$(x_1, x_2) = \begin{cases} \tau s + \cdots & \text{for } h \leqslant 0, \\ \tau s(\tfrac{1}{2}h)^{1/2} + \cdots & \text{for } h > 0 \end{cases}$$

and (only for $h < 0$) from the equation of the sphere

$$x_0^2 + x_1^2 + x_2^2 = \frac{2x_0}{\sqrt{2|h|}}, \qquad x_0 = \sqrt{\frac{2}{|h|}} - s^2\sqrt{\frac{|h|}{2}} + \cdots .$$

Hence by (14)–(16) and (11) we obtain $r = s^2\tau/2 + \cdots$ and $p = 2\tau/s + \cdots$. Furthermore, according to (8), $dt/ds = |r| = s^2/2 + \cdots$, and so

$$t - t_0 = \frac{s^3}{6} + \cdots . \tag{18}$$

Thus r, p and t are analytic functions of the "uniformizing variable" s, while the velocity $p = dr/dt$ has, with respect to this variable, a simple pole at the point of collision, while r and t remain holomorphic. As functions of t the coordinates and velocity have an algebraic singularity (branch point):

$$r = \sqrt[3]{\frac{9}{2}}\,\tau(t - t_0)^{2/3} + \cdots ,$$

$$p = \sqrt[3]{\frac{4}{3}}\,\tau(t - t_0)^{-1/3} + \cdots . \tag{19}$$

Beyond the instant of collision these functions can be continued (preserving reality) in a unique manner.

We now consider how the correspondence between the phase flow of the Kepler problem and the geodesic flow on $\widetilde{M}_h$ is connected with the presence of supplementary first integrals, which are absent in the case of an arbitrary central force (7). The surfaces $\widetilde{M}_h$ are homogeneous in higher powers: on each of them there acts a three-parameter group of motions; that is to say, mappings which do not alter the metric, and so preserve the geodesic flow. By Noether's theorem, to every one-parameter subgroup of this group (more precisely, to its infinitesimal generator, i.e. element of the Lie algebra) there corresponds a first integral. It is known that geodesics parametrized by arc-length are the extremals of the action integral ([7], §12)

$$\int_{s_0}^{s_1} \sum_{i,j} g_{ij}(p)\,\dot{p}_i\dot{p}_j\,ds$$

(the dot indicates differentiation with respect to s). Mappings which do not change the metric leave this integral invariant also. If $X = \{X_i(p)\}$ is a vector field generating a one-parameter group of motions (i.e. being its infinitesimal generator), then the corresponding first integral has the following form ([2], §16):

$$\sum_i g_{ij}(p)\dot{p}_i X_j(p) \equiv \text{Const.}$$

Substituting the values of the coefficients of the metric (13) and using (11), we map this expression into the form

$$(-r_1)X_1 + (-r_2)X_2 = \text{Const} \tag{20}$$

(of course the minus signs may be dropped here without loss; they have been retained solely for the sake of agreement with classical formulas).

The Lie algebra of the group of motions of the surface $\widetilde{M}_h$ is three-dimensional. The vector fields corresponding to the three one-parameter subgroups are easily calculated. These will be:

a) $X_1 = -p_2, X_2 = p_1$

(rotation of the (p_1, p_2)-plane about the origin);

b) $X_1 = (p_1^2 - p_2^2)/2 - h, X_2 = p_1 p_2$

(motions leaving in place the geodesic $p_2 = 0$); and

c) $X_1 = p_1 p_2, X_2 = (p_2^2 - p_1^2)/2 - h$

(motions leaving in place the geodesic $p_1 = 0$).

In the coordinates x_i, a) also corresponds to a rotation of the plane about the origin (about the axis $0x_0$ for $h < 0$). As for b) and c), they correspond for $h < 0$ to a rotation of the sphere about two mutually perpendicular horizontal axes passing through its center, with an angular velocity $(2|h|)^{1/2}$. For $h = 0$ the fields b) and c) generate parallel shifts of the (x_1, x_2)-plane (this agrees also with the limit passage as $h \longrightarrow 0$, under which the radius $(2|h|)^{-1/2}$ of the sphere tends to infinity, but the linear velocity of its "south pole", where $x_1 = x_2 = x_0 = 0$, remains constant). For $h > 0$ the fields b) and c) generate subgroups of noneuclidean parallel shifts along the coordinate axes $x_2 = 0$ and $x_1 = 0$.

Substituting the expressions for X_i in (20), we find the following first integrals:

a) $r_1 p_2 - r_2 p_1 = C$

(this is the classical integral of angular momentum; see (4)),

$$\text{b) } hr_1 - r_1 \frac{p_1^2 - p_2^2}{2} - r_2 p_1 p_2 = l_1,$$

$$\text{c) } hr_2 - r_1 p_1 p_2 + r_2 \frac{p_1^2 - p_2^2}{2} = l_2.$$

The last two expressions may be transformed, using the already-found integral of angular momentum and the integral of energy $2h = p_1^2 + p_2^2 - 2/\sqrt{r_1^2 + r_2^2}$:

$$l_1 = Cp_2 - \frac{r_1}{\sqrt{r_1^2 + r_2^2}}, \qquad l_2 = -Cp_1 - \frac{r_2}{\sqrt{r_1^2 + r_2^2}},$$

from which it is apparent that l_1 and l_2 are none other than components of the so-called "invariant vector of Laplace". In $\mathbf{R}^3$ this vector is written thus:

$$\vec{l} = \vec{p} \times [\vec{r} \times \vec{p}] - \vec{r}/|\vec{r}|,$$

and, in conformity with the plane case, where

$$\vec{r} = (r_1, r_2, 0), \quad \vec{p} = (p_1, p_2, 0), \quad \vec{r} \times \vec{p} = (0, 0, C),$$

we obtain the above expressions for the components of the vector $\vec{l} = (l_1, l_2, 0)$.

The invariant vector $\vec{l}$ constitutes in fact the supplementary first integral which is specific for the Newtonian law of interaction ([4], §15).

The integrals obtained are connected by the relation

$$l_1^2 + l_2^2 = 2hC^2 + 1,$$

which is natural, since the phase space of the plane Kepler problem is four-dimensional. And so not more than three first integrals can be independent.

The geometrical interpretation of the Kepler problem makes it possible to give a lucid discussion of the elements required by celestial mechanics: the mean, true and eccentric anomalies, and so on (see [44]). We remark further that $(l_1^2 + l_2^2)^{1/2}$ is equal to the eccentricity of the orbit.

§3. The three-body problem. Analytic properties of solutions

The problem of the motion of three material points under the action of Newtonian laws of mutual attraction–the three-body problem–has attained wide recognition in mathematics, mechanics and astronomy. It is sufficient to glance at the chapters devoted to this problem in the books of Whittaker [12], G. D. Birkhoff [1], Siegel [5], and Siegel and Moser [6], and the above-mentioned articles of Arnol'd [18] and Smale [31], to be convinced of the richness and fruitfulness of the set of ideas which in one way or another owe their origin to it.

The three-body problem is described by a system of differential equations (1) (of §1): to it corresponds a phase-flow in 18-dimensional phase space. Choosing an inertial coordinate system with origin at the center of mass, we obtain (2), so that one can reduce the problem to 6 degrees of freedom in a 12-dimensional phase space M^{12}. After this, the problem still has 4 algebraic first integrals (3) and (4), whose level surfaces stratify M^{12} into eight-dimensional manifolds, whose topology is studied in [31]. By the theorem of Bruns (see [12]), in M^{12} there do not exist algebraic first integrals distinct from the four classical ones; and Poincaré [8] showed that there do not even exist single-valued analytic first integrals which are analytically dependent also on the masses of the bodies. This suggests the idea of an intricate behavior of the trajectories in the phase space.

It is quite natural, if one is unsuccessful in finding the general solution, to attempt to find particular solutions of the three-body problem for which an integration is possible, for example taking into account considerations of symmetry. It is easy to see that a system of gravitating material points cannot have a state of static equilibrium. Lagrange and Euler succeeded, however, in showing that dynamical equilibrium is possible; three bodies are situated in points with unchanging coordinates, but in a non-inertial, uniformly rotating coordinate system. In other words, each of the bodies executes uniform circular motion about the common center of mass, with the same angular velocity.

For the Lagrange solution the bodies P_1, P_2 and P_3 are then located at the vertices of an equilateral triangle. In the solar system such a configuration is formed fairly accurately by the Sun, Jupiter, and each of the asteroids of the so-called Trojan group.

For the Euler solution the triangle $P_1P_2P_3$ degenerates, and all three bodies lie on a single line. If the masses of the bodies are distinct, there are three classes of such solutions, depending on which of the three bodies lies between the other two.

There are also other special solutions, in which the configuration of the three bodies remains similar to itself through all time [11]. In some, which may be called "Lagrangian" [67], the triangle $P_1P_2P_3$ is equilateral; in others, "Eulerian" [68], all three bodies lie on a straight line; each of the bodies P_i moves around the common center of mass along some conic section, which may degenerate into a straight line. We see later that these solutions are to be encountered, in an unexpected manner, in two different aspects of the qualitative analysis of the three-body problem.

The right-hand sides of the system of differential equations

$$\ddot{r}_i = \sum_{j \neq i} \frac{\gamma m_j r_{ji}}{|r_{ji}|^3}, \quad i, j = 1, \ldots, n, \tag{1}$$

are real analytic functions of the coordinates, and by Cauchy's theorem the solution is uniquely determined by the initial values, and is an analytic function of the time t, at least locally. This function can be continued analytically along the whole time axis, so long as we do not meet singularities along the way.

We saw above that in the case $n = 2$ this possibility can actually be realized for all solutions with zero angular momentum, since for each of them the bodies have at least one collision, i.e. there is an instant t_0 (or instants, if $h < 0$), such that as $t \longrightarrow t_0$ the distance $r_{12} \longrightarrow 0$. This singularity turns out to be algebraic (see (18) and (19)), and after transition to a new independent variable

$$s = A(t - t_0)^{1/3} + \cdots \tag{21}$$

the coordinates of the bodies become analytic functions of s in some neighborhood of $s = 0$, while the velocities have a pole of the first order. The solution has a unique real analytic continuation through the point $t = t_0$, corresponding to the real branch of the cube root in (21). Solutions of the two-body problem have no other singularities on the real axis.

In the case $n = 3$, in addition to pairwise collisions, there is the possibility of a triple collision. This exhausts all possibilities, as the following theorem shows.

THEOREM OF PAINLEVÉ ([45], [3], [5]). *If a solution of the three-body problem is an analytic function of t in the interval $(0, t_0)$, but ceases to be so for $t = t_0 \neq \infty$, then as $t \longrightarrow t_0 - 0$ either all distances $r_{ij} \longrightarrow 0$, or else only one of them tends to 0 and the other two tend to a finite positive limit.*

If the first possibility holds, then by the convention (2) concerning the center of mass all the $r_i \longrightarrow 0$ as $t \longrightarrow t_0 - 0$. It is important to note that in the case of a pairwise collision (the second possibility) the radius-vectors of all three bodies also tend to finite limits $r_i(t) \longrightarrow r_i(t_0)$, and if, for example, $r_{12} \longrightarrow 0$ then $r_1(t_0) = r_2(t_0)$.

Unfortunately, one has not succeeded so far in proving an analogous theorem for $n > 3$. From Painlevé's argument it follows only that for $t \longrightarrow t_0 - 0$ the least of the r_{ij} tends to 0. In 1908, von Zeipel formulated a conditional assertion: if as $t \longrightarrow t_0 - 0$ all the distances r_{ij} remain bounded, then the bodies $P_1, \ldots, P_n$ may be sorted into a certain number of groups such that as $t \longrightarrow t_0 - 0$ the bodies in one group collide at some definite point in space. The proof of this assertion was subjected to criticism by Wintner [11], and rehabilitated in [49]. The alternative possibility to that of von Zeipel's assertion, namely that as $t \longrightarrow t_0 \neq \infty$ some of the distances r_{ij} are unbounded, appears unlikely, particularly from the physical point of view, but remains so far without disproof.

We return to the three-body problem. A triple collision is in a certain sense a highly unlikely event; and, what is particularly important, one can give a simple sufficient condition for its absence. In fact:

THEOREM. *If in the n-body problem all the* $r_{ij} \longrightarrow 0$ *as* $t \longrightarrow t_0$, *then the moment of momentum of the system vanishes.*

This theorem was proved long ago by F. A. Sludskiĭ [30]; it was apparently known to Weierstrass and independently proved by Sundman in his remarkable paper [51] on the regularization of the three-body problem (see also [3] and [5]).

Juxtaposing the theorems of Painlevé and Sludskiĭ, we conclude that solutions of the three-body problem with nonzero moment of momentum can have only singularities of the pairwise collision type. For the regularization of these singularities Sundman introduced a new independent variable s with the aid of the formula

$$t = \int \prod_{i<j} (1 - e^{-r_{ij}/l}) ds, \tag{22}$$

where l is a suitable constant, dependent only on the masses and the initial values of the coordinates and the velocities of the bodies. The distinction between the variables t and s becomes essential only when one of the distances r_{ij} vanishes, and obviously in this case they are related to one another roughly as in formula (8) in §2. Thus the substitution (22) is of a universal nature and makes it possible to regularize all pairwise collisions, irrespective of which pair collides.

THEOREM OF SUNDMAN. *If a solution of the three-body problem has nonzero moment of momentum, then there exist* l *and* δ *such that in the strip* $|\mathrm{Im}\, s| < \delta$ *the coordinates of all three bodies, the pairwise distances between them, and the time* t *are analytic functions of the variable* s, *which is connected with the time* t *by the relation* (22). *In the same strip the velocities of the bodies can have on the real s-axis only poles of the first order.*

For the proof of this theorem Sundman had to overcome two basic difficulties. The first is connected with the regularization of individual pairwise collisions. If for

$t = t_0$ there is, say, a collision between P_1 and P_2, then around this instant the force of interaction of these bodies will be much greater than the forces of their interaction with P_3. Hence the mappings which lead to the regularization must here be essentially the same as those given in §2 for the case when other bodies are absent, except for those colliding. The singularity is again algebraic; as in (19), the coordinates have a single branch point of the third order, and there exists a unique real analytic continuation past the instant of collision. (Recently Sperling [49] extended this result to the problem of several bodies in the case when it is known beforehand that the singularity has the character of pairwise collisions, possibly of several pairs of bodies.)

The second difficulty consists in proving a uniform bound for the radii of convergence of the series which represent the solution, i.e. the analyticity of the solution in a strip of uniform width surrounding the real axis. The deduction of the requisite bounds for the perimeter of the triangle formed by the three bodies and the magnitude of the velocity of the body not taking part in the collision requires delicate and laborious arguments and takes up considerable space in Sundman's work.

A regularization of a triple collision, similar to that done by Sundman for the case of pairwise collisions, turns out to be impossible. The analytic properties of solutions of the three-body problem near a point of triple collision have been investigated by a number of authors (see [50], [37] and [32]). This question was studied most fully, it seems, by Siegel, an exposition of whose results is given in the new edition [6] of his well-known "Lectures on celestial mechanics", supplemented by J. K. Moser. Without pausing for a detailed account, we mention only two interesting facts.

In the first place (and herein lies the assertion of the impossibility of regularization), a solution of the three-body problem, analytic on some interval $(0, t_0)$ and having at $t = t_0$ a singularity of triple collision type, does not have in general a real analytic continuation for $t > t_0$ (the expressions for the coordinates contain $(t - t_0)^\lambda$, and the exponent λ may be irrational).

Secondly, under approach to the instant t_0 of triple collision at the origin a limiting configuration of the three bodies exists, and in fact the following limits exist:

$$\lim_{t \to t_0} \frac{r_i(t)}{(t - t_0)^{2/3}} = R_i .$$

Here it turns out that the points R_i are either located at the vertices of an equilateral triangle, or else lie on a straight line. Thus in the neighborhood of a triple collision the motion of the three bodies is close either to the Lagrange or to the Euler particular solution, which emphasises the important role of these solutions for the quantitative analysis of the whole problem in the large.

According to [37], the collection of all trajectories in the phase space M^{12}, on which the triple collision takes place, forms four submanifolds: one is seven-dimensional, corresponding to motions with Lagrangian asymptotics, and three are five-dimensional, corresponding to motions with Eulerian asymptotics (we recall that there are three types of Eulerian motion, depending on which of the three bodies lies between the other two).

All these manifolds lie on a nine-dimensional algebraic submanifold in M^{12}, on which the integral of moment of momentum (4) in §1 vanishes.

As regards the trajectories with pairwise collisions, it is easily seen that their manifold is ten-dimensional. It is unknown how badly such a manifold can be situated in M^{12}, for example whether it may not be everywhere dense in some open subset of M^{12}.

Concluding this section, we mention that if we map the strip $|\mathrm{Im}\ s| < \delta$ conformally onto the unit disc, we obtain a solution which by Sundman's theorem is analytic in this disc, and so can be expanded in a convergent power series. From the theoretical point of view, this series will be suitable for calculating the coordinates of the bodies on the whole time axis. The French astronomer Belorizky estimated the number of terms of the series which would make possible the calculation with a degree of accuracy sufficient for the needs of practical astronomy. It turned out that for this it would be necessary to take $10^{8000000}$ terms!

§4. Final motions in the three-body problem

The investigation of the qualitative properties of solutions of the three-body problem have proceeded for a relatively long time in the direction of studying final motions, i.e. the behavior of the solutions as $t \longrightarrow \infty$. The long-standing and steadfast interest which has attached to these questions for specialists and nonspecialists alike is fully explicable. One is dealing here with problems which arise in the area where mathematics and mechanics border on philosophy: the origin and fate of the solar system, the evolution of stellar clusters, and so on.

One of the first results in this area was obtained by Laplace and Lagrange, who proved that to the first approximation (i.e. if one neglects quantities containing squares of the ratios of the masses of the planets to the mass of the sun), the motion of the planets may be described by purely trigonometric series [9]. This result was interpreted as a proof of the stability of the solar system; with this is linked the origin of the notion "stability in the sense of Lagrange" in the general theory of dynamical systems. Unfortunately, hopes for the validity of an analogous assertion in higher approximations of perturbation theory were not substantiated, and at the present time the question of the stability, both of the real solar system, and in general of planetary systems of gravitating points, remains open; we will return to this question later.

The simplest and at the same time most graphic characteristic of the motion is the asymptotic behavior as $t \longrightarrow \infty$ of the pairwise distances between the bodies.

The Newtonian character of the force of interaction gives rise to two typical forms of asymptotic behavior for escape to infinity; the hyperbolic $O(t)$ and the parabolic $O(t^{2/3})$. In the one-dimensional Kepler case this is obvious: from the energy-integral

$$\frac{\rho^2}{2} - \frac{1}{\rho} = h$$

we find that

$$\rho = \sqrt{2h} \cdot t + O(\ln t) \qquad \text{for } h > 0;$$

$$\rho = \left(\frac{2}{3}t\right)^{2/3} \qquad \text{for } h = 0;$$

$$\rho \leqslant 1/|h| \qquad \text{for } h < 0.$$

We have the same three cases also in the two-body problem. Depending on the sign of the total energy constant h (the energy of relative motion of these two bodies, if the center of mass is not supposed at rest), the motions may be divided into

H, hyperbolic: $\rho_{12} \longrightarrow \infty, \rho_{12}/t \longrightarrow C > 0$;

P, parabolic: $\rho_{12} \longrightarrow \infty$, $\rho_{12}/t^{2/3} \longrightarrow C > 0$;

B, bounded (or elliptic): $\rho_{12} < C < \infty$.

(Here and in the sequel $\rho_{ij} = |r_{ij}|$ is the distance between the bodies P_i and P_j; C is some constant, not necessarily the same in all formulas.)

In the case of an arbitrary number of bodies the situation turns out to be much more complicated. However, Saari was able to prove [48] a certain qualified assertion, according to which three of the above types of behavior as $t \longrightarrow \infty$ of the pairwise distances are basic. According to Saari, all n of the bodies can be divided up into clusters, within which the pairwise distances are bounded; the clusters are grouped into subsystems such that within one subsystem the distances between the clusters is of the order $t^{2/3}$; finally, the distances of the centers of mass of the subsystems from the common center of mass of the whole system are of order t. This beautiful result was obtained, however, assuming a number of conditions which do not admit effective verification, the most important of which is the absence of oscillating triples (ijk), where $\rho_{ij} \longrightarrow \infty$ but the ratio ρ_{ij}/ρ_{ik}, even if not bounded, does not tend to infinity.

We return to the three-body problem. Here we have in all three pairwise distances, which for brevity we denote by $\rho_1 = \rho_{23}$, $\rho_2 = \rho_{13}$ and $\rho_3 = \rho_{12}$. The quantities ρ_i are connected by the triangle inequality, which reduces the number of logically possible combinations of their asymptotics. Back in 1922 the French astronomer and mathematician Jean Chazy [38] described all possible types of final motion in the three-body problem. According to Chazy, the whole phase space is divided into the following subsets:

1) H, hyperbolic motions: $\rho_i \longrightarrow \infty$ and $\rho_i/t \longrightarrow C_i > 0$;

2) HP_k, hyperbolic-parabolic motions:

$$\rho_i \longrightarrow \infty; \quad \rho_k/t^{2/3} \longrightarrow C > 0; \quad \rho_i/t \longrightarrow C_i > 0, \quad i \neq k;$$

3) P, parabolic motions:

$$\rho_i \longrightarrow \infty; \quad \rho_i/t^{2/3} \longrightarrow C_i > 0.$$

In all three cases the system of three bodies (a triple star) breaks up into separate constituent components; the difference is only in the speed with which the components move away from each other. For example, for a motion from HP_1 the system falls into two subsystems: one consists of the one body P_1, and the other of the bodies P_2 and P_3 which are moving away from each other in parabolic fashion ($\sim t^{2/3}$), while the subsystems are diverging hyperbolically ($\sim t$).

4) HE_k, hyperbolic-elliptic motions:

$$\rho_k < C; \quad \rho_i \to \infty; \quad \rho_i/t \to C_i > 0, \quad i \neq k;$$

5) PE_k, parabolic-elliptic motions:

$$\rho_k < C; \quad \rho_i \to \infty; \quad \rho_i/t^{2/3} \to C_i > 0, \quad i \neq k.$$

In these two cases the system decomposes into a double star and a single body, the distance between which tends to infinity hyperbolically or parabolically; each case may be divided into three subcases, depending on which of the three bodies is moving away from the other two. For example, for motions PE_2, the distance between P_1 and P_3 is bounded, while P_2 is moving away from them unboundedly and parabolically.

6) B, bounded motions: all $\rho_i < C < \infty$;

7) OS, oscillatory motions: at least one of the ρ_i is unbounded, but does not tend to infinity.

The existence of motions of the types H, P and B was known already to Chazy; examples are furnished, say, by Lagrangian solutions, in which the triangle $P_1P_2P_3$ is equilateral (of varying size). He did not doubt the existence of motions of hyperbolic-elliptic type (although a more or less rigorous proof was given, apparently, only by Birkhoff in [1]), because they are easily constructed on the basis of combination of two Keplerian motions. The cases HP_k and PE_k are limiting cases.

As regards the class of oscillatory motions, these were introduced by Chazy from purely logical considerations, and for a long time the existence of such motions remained in doubt. Only in 1959 was it shown by K. A. Sitnikov [29] that $\mathrm{OS} \neq \varnothing$.

The mutual disposition of the subsets 1)–7), as conceived by Chazy, is represented in Figure 4. It follows from the definitions and formula (3) of §1 that H and HP_k lie in a region where the energy constant $h > 0$, P is on the hypersurface $h = 0$, and PE_k is in the region where $h < 0$. Chazy showed that B and OS also lie in the region where $h < 0$. Hyperbolic-elliptic motions HE_k are possible for any sign of h.

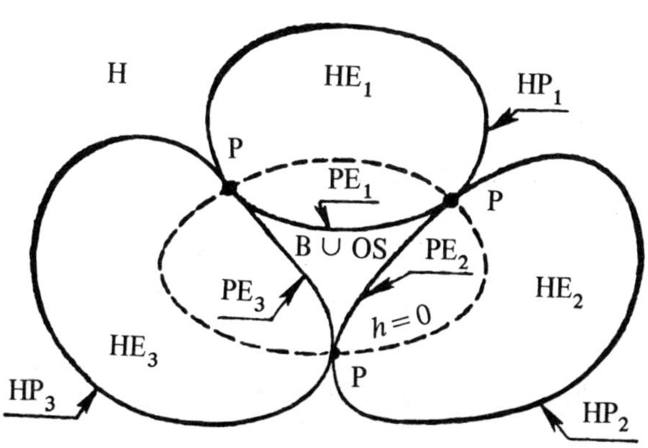

FIGURE 4

For motions belonging to the set

$$A_k = \mathrm{H} \cup \mathrm{HP}_k \cup \mathrm{HE}_k \cup \mathrm{P} \cup \mathrm{PE}_k,$$

it is convenient to use the coordinates of Jacobi. For A_1 they are introduced as in Figure 5: $r = r_{23} = r_2 - r_3$, and R is a vector beginning at the center of mass of the

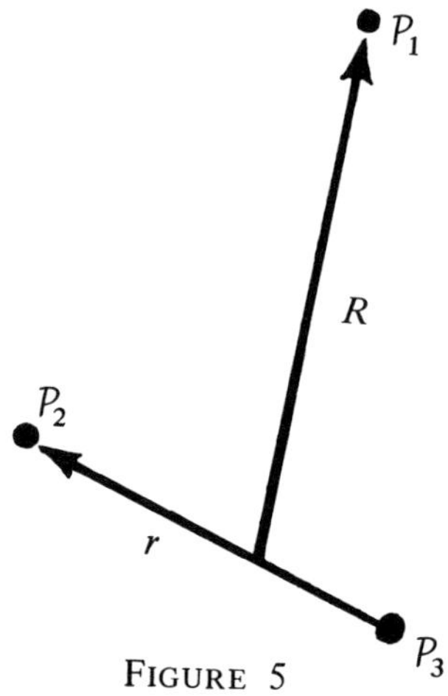

FIGURE 5

bodies P_2 and P_3 and ending in P_1 (for A_2 and A_3 the indices undergo the corresponding changes). In Jacobi coordinates the equations of motion have the form

$$\ddot{r} = -\frac{\mu_1 r}{|r|^3} + \Phi_1(r, R), \qquad \ddot{R} = -\frac{\mu_2 R}{|R|^3} + \Phi_2(r, R), \tag{23}$$

where the μ_i are certain constants, dependent only on the masses m_k, and the functions $\Phi_i = O(|r|/|R|^3)$. For motions of the set A_1 the ratio $|r|/|R|^3$ tends to 0 as $t \longrightarrow \infty$ so rapidly that the system (23) decomposes asymptotically into two Keplerian subsystems. Chazy proved that the elements determining the momentary Keplerian orbits in these two subsystems tend as $t \longrightarrow \infty$ to definite limits. In particular, there exist the limiting relative energies

$$h_r = \lim_{t \to +\infty} \left(\frac{\dot{r}^2}{2} - \frac{\mu_1}{|r|} \right), \qquad h_R = \lim_{t \to +\infty} \left(\frac{\dot{R}^2}{2} - \frac{\mu_2}{|R|} \right), \tag{24}$$

and the membership of motions in the various final types may be characterized in terms of these quantities thus:

$$\mathrm{H} = \{h_r > 0, h_R > 0\}, \quad \mathrm{HP}_1 = \{h_r = 0, h_R > 0\},$$

$$\mathrm{HE}_1 = \{h_r < 0, h_R > 0\}, \quad \mathrm{P} = \{h_r = 0, h_R = 0\},$$

$$\mathrm{PE}_1 = \{h_r < 0, h_R = 0\}.$$

Both the Keplerian systems into which the system (23) decomposes asymptotically can be fully integrated, so that the following conjecture appears extremely plausible:

CONJECTURE. In the region

$$A = \operatorname{Int}(A_1 \cup A_2 \cup A_3) = \mathrm{H} \cup \bigcup_k [\mathrm{HE}_k \cup \mathrm{HP}_k] \tag{25}$$

the three-body problem has a complete set of single-valued analytic first integrals.

Chazy considered this assertion valid, and proposed constructing the necessary integrals from limiting values of elements of the momentary Keplerian orbits in subsystems. However, his arguments, given in [38] and [40], do not seem exhaustive to me. If nevertheless it is true, it would be interesting to study the behavior of the integrals near the boundary of (25), which is extremely complicated.

§ 5. Evolution of the system. The problem of capture

As already stated, in the Kepler problem or in the two-body problem leading to it the final type of the motion is determined by the energy constant, and remains the same for $t \longrightarrow +\infty$ and for $t \longrightarrow -\infty$. In his memoirs [39] and [40] Chazy formulated an analogous assertion for the three-body problem also, and for a fairly long time the mathematical, and especially the astronomical world was convinced that such a remarkable symmetry really holds good. Also in its favor was the fact that in all cases when it was possible to obtain exact results (not using numerical integration), the behavior of solutions as $t \longrightarrow \pm\infty$ turned out to be the same. Apart from particular solutions (Euler's, Lagrange's, Poincaré's periodic solutions, and so on), one may cite here also the theorems of Birkhoff [1], Chapter 9, from which it followed that the sets([4]) $\mathrm{H}^+ \cap \mathrm{H}^-$ and $\mathrm{HE}_k^+ \cap \mathrm{HE}_k^-$ contain interior points. In these theorems, apparently, there appears for the first time what subsequently came to be called criteria for the final time of a motion. These criteria have the following form. A subset (generally a region) is determined in phase space by a system of inequalities, and a theorem is proved: if the initial conditions belong to the region so defined, then the final type of the motion is such and such.

A certain dissonance in the symmetry picture of Chazy was introduced by examples of L. Becker [36], obtained by numerical integration. The initial conditions and intervals of integration were chosen so that it seemed obvious that the motion belonged to the class $\mathrm{HE}_1^- \cap \mathrm{HE}_2^+$ if, furthermore, one could verify that as $t \longrightarrow \pm\infty$ it was really hyperbolic-elliptic. The change in the class of motion contradicted the results of [39] and [40]. However, Chazy attributed this to errors in the numerical integration and to the impossibility of following the behavior of the solution over an infinite time-interval, and even remarked in this connection that precise mathematical analysis demonstrates its superiority here over approximate numerical methods of investigation.

Interest in the qualitative properties of solutions of the n-body problem was an important question in the 40's, in connection with the cosmological theory of O. Ju. Šmidt. According to this theory, the planets of the solar system had their origin in a

([4]) The indices ± here and in the sequel indicate final types of motion as $t \longrightarrow \pm\infty$, respectively.

meteoritic dust cloud encircling the Sun that had been captured by the Sun on passage through a haze of particulate matter. If one confines attention to purely gravitational interaction, a capture of this kind means a change in the final type of motion in going from $t = -\infty$ to $t = +\infty$. Although Chazy's conclusions related only to the three-body problem, and it is not possible to use them for arguments against the possibility of capture in the n-body problem, nevertheless this evoked justified scepticism in regard to Šmidt's theory. To support his hypothesis, Šmidt [34] constructed by numerical integration a counterexample to the basic assertion of [40]. In this example, three stars independent in the past (a motion of type H^-) form a stable subsystem (a double star), while the third star again goes off to infinity (motion of type HE_3^+). This phenomenon may be termed partial capture.

Šmidt's example, based on numerical integration, was open to criticism from the same points of view as the examples in [36]. One of the objections put forward was dealt with by Šmidt's collaborator G. F. Hil′mi [33], [13], who constructed criteria for hyperbolic and hyperbolic-elliptic motions, in the sense indicated above. The resulting situation is shown schematically in Figure 6.

The arc AB represents a trajectory in the phase space, obtained by numerical integration. At the points A and B the inequalities of Hil′mi's criteria are fulfilled, so that after the point B the trajectory remains forever in the region HE_3^+ (i.e. the motion is hyperbolic-elliptic), while in motion from the point A in the sense of decreasing time we remain in the region H^- (hyperbolic motion).

However there remains an objection connected with errors in the numerical integration which is difficult to allow for. A thorough check showed that the original calculations contained an error, in fact at a point where the Hil′mi criterion did not yet hold (in Figure 6 schematically at the point C) [26]. Fortunately, G. A. Merman put

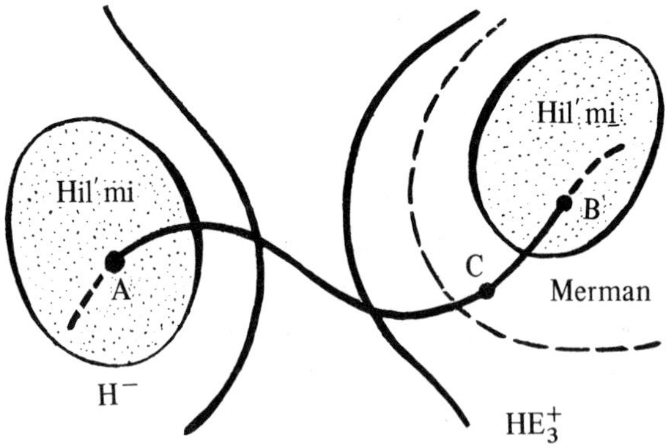

FIGURE 6

forward [21] a criterion more general than Hil′mi's; at the point C Merman's criterion is fulfilled, and this saved Šmidt's example (see [24]). All this has, however, by this time lost some of its point, since in 1953 K. A. Sitnikov constructed [28] an example of partial capture ($H^- \cap HE^+$) by a purely qualitative route, without using numerical integration. In principle Sitnikov's example is similar to Šmidt's, but instead of computing the arc of the trajectory AB, one proves the existence of an arc going from the region of applicability of one criterion to the region of application of the other.

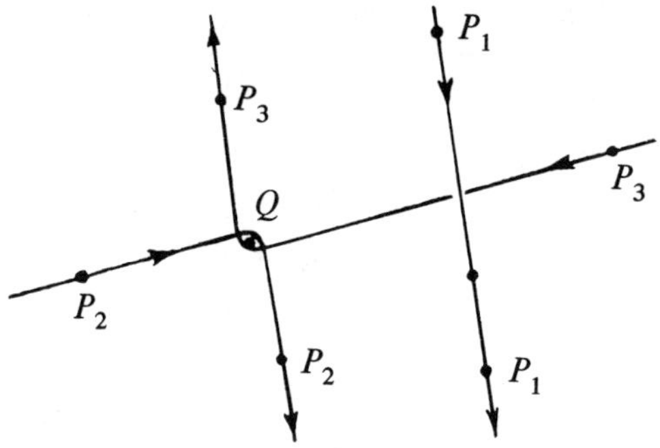

FIGURE 7

This may be done as follows. Let the bodies P_2 and P_3 move with very great velocities and near $t = 0$ pass close to each other in the vicinity of the point Q (Figure 7). Their relative motion is here hyperbolic, and by a suitable choice of the velocities and by "aiming the separation" the angle between the asymptotics of the hyperbola may be made sufficiently large and fixed. After the approach, the motion of these same bodies remains almost rectilinear, but its direction changes (only near the point Q are the hyperbolas really "curves", while far from this point they almost coincide with their asymptotes). We now assume that the initial velocity of P_1 is given so that it is almost equal to the velocity of P_2 after the approach. Then for $t < 0$ the velocities of the three bodies are different in direction and, if they are sufficiently large, will satisfy the criterion for hyperbolic motion. For positive t, on the other hand, P_3 will rapidly leave the pair P_1, P_2 which, in turn, will not be able to diverge from each other, and at a suitable moment the criterion for hyperbolic-elliptic motion will be satisfied. This example, with minor variations as to details, lies at the base both of Sitnikov's and Šmidt's examples.

The discussion of the problem of capture touched off a long series of investigations, devoted both to a critical analysis of Chazy's work, and also to the whole area of problems connected with the final types of motions (in addition to the above-mentioned, see [22], [23], [27], [16], and others). Some of the relevant results are reflected in Tables 1 and 2. Each box corresponds to one of the logically possible combinations of the basic types of final motion (H, HE_k, B, OS) in the past and future, and thus describes some type of evolution of the system. Authors are cited, with dates, in which the corresponding types were found; however, this information is sometimes rather involved. Also indicated is the Lebesgue measure of the corresponding set in the manifold M^{12}. It should be borne in mind that because of symmetry in time each investigation, dealing with one of the boxes, applies in the same way to the box symmetrically placed with respect to the leading diagonal. Thus, the existence of examples of partial capture $H^- \cap HE_k^+$ means at the same time the existence of examples of total dispersion $HE_k^- \cap H^+$.

TABLE 1

$h > 0$		$t \longrightarrow +\infty$: H^+	$t \longrightarrow +\infty$: HE_i^+
$t \longrightarrow -\infty$	H^-	Lagrange, 1772 (special examples) Chazy, 1922 measure > 0	PARTIAL CAPTURE Measure > 0 Šmidt (numerical example), 1947 Sitnikov (qualitative methods), 1953
$t \longrightarrow -\infty$	HE_j^-	TOTAL DISPERSAL Measure > 0	$i = j$, Measure > 0 Birkhoff, 1927
$t \longrightarrow -\infty$	HE_j^-		$i \neq j$, EXCHANGE, Measure > 0 Becker (numerical examples), 1920 Alekseev (qualitative methods), 1956

TABLE 2

$h < 0$		$t \longrightarrow +\infty$: HE_i^+	$t \longrightarrow +\infty$: B^+	$t \longrightarrow +\infty$: OS^+
$t \longrightarrow -\infty$	HE_j^-	$i = j$, Measure > 0 Birkhoff, 1927 ------------- EXCHANGE $i \neq j$, Measure > 0 Becker, 1920 (numerical examples) Alekseev, 1956 (qualitative methods)	TOTAL CAPTURE Measure $= 0$ Chazy, 1929; Merman, 1954; Littlewood, 1952; Alekseev, $\neq \varnothing$, 1968	Measure $= 0$ Chazy, 1929; Merman, 1954; Alekseev, $\neq \varnothing$, 1968
$t \longrightarrow -\infty$	B^-	PARTIAL DISPERSAL $\neq \varnothing$ Measure $= 0$	Euler, 1772; Lagrange, 1772; Poincaré, 1892 (special examples); Arnol'd, 1963 Measure > 0	Measure $= 0$ Littlewood, 1952; Alekseev, $\neq \varnothing$, 1968
$t \longrightarrow -\infty$	OS^-	Measure $= 0$ $\neq \varnothing$	Measure $= 0$ $\neq \varnothing$	Sitnikov, $\neq \varnothing$, 1959 Measure $=$?

In the case $h > 0$ (Table 1), despite the memoir [40], all the logically possible types of evolution have turned out to be realizable.

Since the sets $\mathrm{H}^{\pm}$ and $\mathrm{H}_k^{\pm}$ are open, this automatically ensures a positive probability (a measure > 0) for each type. For example, from the existence of a particular Lagrange solution, in which all three bodies move along hyperbolas around a common center of gravity, forming the whole time an equilateral triangle, it follows that $\mathrm{H}^- \cap \mathrm{H}^+$ is nonempty, and so has positive measure. In the corresponding box in Table 1 we have written "Lagrange, 1772" and "Chazy, 1922", although with no less justification we could have written "Birkhoff, 1927", which I in fact did in [35] (the openness of the sets H^+ and H^- was known to Chazy, and he knew of Lagrange's example; however in [38] Chazy studied only the one-sided behavior of solutions).

In the case $h < 0$ the situration is significantly more complicated than for $h > 0$. First of all, the sets $\mathrm{PE}_k^{\pm}$, though submanifolds of codimension 1, are certainly badly imbedded in M^{12}, so that the picture turns out to be not so simple as in Figure 4. It is not clear whether these manifolds are analytic. The following conjecture, originating with A. N. Kolmogorov, appears very plausible.

CONJECTURE. Almost every point $p \in \mathrm{PE}_k^+$ is attainable from HE_k^+, and has in M^{12} a neighborhood U, diffeomorphic to $I \times D^{11}$, such that

$$U \cap \mathrm{HE}_j^+ \simeq A_j \times D^{11}, \quad U \cap \mathrm{PE}_j^+ \simeq B_j \times D^{11},$$

$$U \cap (\mathrm{B}^+ \cup \mathrm{OS}^+) \simeq C \times D^{11},$$

where D^{11} is an eleven-dimensional disc, and $I = (-\epsilon, \epsilon)$, A_j, B_j, $C \subset I$, where each of the A_j, $j = 1, 2, 3$, consists of a countable number of intervals, B_j is countable and formed by the ends of intervals corresponding to A_j, and C is zero-dimensional and homeomorphic to the Cantor set.

It is not clear what degree of smoothness the diffeomorphism $U \simeq I \times D^{11}$ can have; also, analyticity cannot be excluded a priori.

Below we will show that the preceding conjecture is valid in a somewhat simplified situation, where considerations of symmetry make it possible to reduce the dimension of the phase space.

The sets $\mathrm{HE}_k^{\pm}$ are open and connected; however, each of them is very "branched" in M^{12}, and individual branches interlace one with another in a very confused way. Birkhoff [1] conceived of $\bigcup_k \mathrm{HE}_k^-$ in the form of three flows, coming from infinity. Continuing this analogy, one should imagine that each of these flows breaks up into a countable number of "rivulets" which, like a system of bloodbearing capillaries, pass through the phase space and are reunited in the three departing flows HE_i^+.

An analysis of Chazy's arguments in [39] showed [22] that, although they did not suffice to prove the equalities $\mathrm{HE}_k^+ = \mathrm{HE}_k^-$, one could nevertheless assert that the sets $\mathrm{HE}^+ = \bigcup_k \mathrm{HE}_k^+$ and $\mathrm{HE}^- = \bigcup_k \mathrm{HE}_k^-$ coincide in the region $h > 0$ to within a set of zero measure in M^{12} ("almost all water carried in by the flows HE_k^-, with the exception of individual streams of total measure zero, is carried away by the flows HE_k^+").

In this connection there arise the questions: do there exist in the region $h < 0$ motions of the type a) $HE_i^- \cap HE_j^+$, $i \neq j$ ("exchange"); b) $HE_k^- \cap B^+$ ("total capture") or $HE_k^- \cap OS^+$?

These questions are by no means of equal value. The construction of an example of exchange could be attempted by following the same principles as in the examples of Šmidt and Sitnikov. Examples from [36] prompted a suitable choice of initial conditions with a certain symmetry. True, the examples of Šmidt and Sitnikov made essential use of large velocities (here motions of the bodies are nearly rectilinear over large stretches, thus making it possible to carry out the corresponding estimates), and so one necessarily found oneself in the region $h > 0$. Nonetheless, almost half the numerical examples from [36] related to the region $h < 0$, which gave hope for the construction of similar examples also by purely qualitative means.

This in fact worked. The general scheme applied in [15] for the construction remains the same as in Figure 6; one need only replace H^- by HE_2^- and use in both cases suitable criteria for hyperbolic-elliptic motion. The construction of the arc AB is based once more on the close passage of two of the three bodies; in place of large velocities one uses as a small parameter the mass of the bodies of the pair which come close.

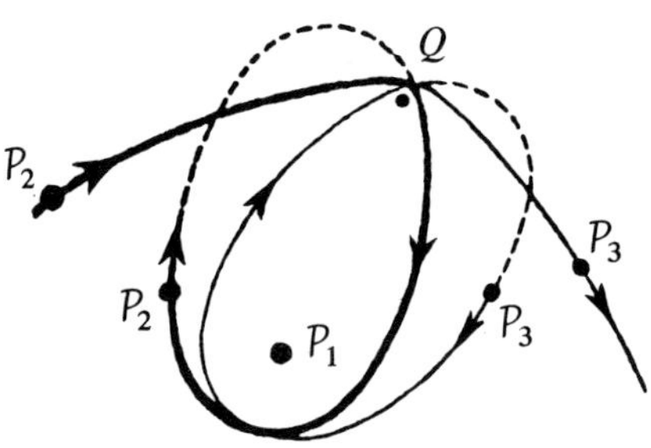

FIGURE 8

Let the mass of P_1 (the "sun") be large in comparison with the masses of P_2 and P_3 (the "planets") and as $t \longrightarrow -\infty$ ("in the remote past") let the motion be hyperbolic-elliptic, of the class HE_2^-. Here P_2 and P_3 move along orbits which respectively are close to a hyperbola and an ellipse with focus in the sun. One can choose these orbits so that near the point Q (Figure 8) the planets approach each other so closely that the strength of their interaction is much bigger than the forces attracting them to the sun (if we assume, for example, that P_2 and P_3 have the mass of the Earth, and that the distance QP_1 is equal to the distance from the Earth to the Sun, then they must pass each other at a distance of the order of 500 km.). After such an approach the velocities of the planets will change sharply, and one may arrange that they interchange their roles as $t \longrightarrow +\infty$, P_3 going to infinity asymptotically along a hyperbola, while P_2 will rotate about P_1 asymptotically along an ellipse. Further details of the construction would constitute a digression from the main theme of our lectures, and we shall not dwell on them (see [16] and [35]).

The second of the above questions turned out to be much more difficult. The set B ∪ OS does not form a region in M^{12}, and so it is not possible to characterize membership in it as was done for H and HE_k. Hence it was necessary to find some new methods which would permit one to follow the behavior of a solution over an infinite time-interval all at once.

With a view to avoiding difficulties connected with the large dimensionality of the phase space, A. N. Kolmogorov proposed in 1954 to study one particular case of the three-body problem, in which symmetry considerations made it possible to reduce the problem to a system with two degrees of freedom. We will study this system in more detail in one of the subsequent parts, and confine ourselves at present to recalling results which could be obtained by this route. First, K. A. Sitnikov [29] showed in 1959 for this example (and so for the general three-body problem) the existence of oscillating motions (OS), which were introduced by Chazy as a purely logical possibility, to be admitted so long as it was not shown that it should be rejected. Strictly speaking, Sitnikov's arguments related solely to one-sided behavior of the solutions, but considerations of symmetry allow one to show the existence of solutions of the type $OS^- \cap OS^+$, as illustrated in Table 2. Kolmogorov showed that underlying Sitnikov's arguments is an extremely simple geometrical construction, and in this connection he made the conjecture about the formation of the boundary of the regions HE_k, mentioned above. The author then succeeded in showing that in the example in question one could apply methods of "symbolic dynamics"; by this means one could prove the nonemptiness of the classes $HE_k^- \cap B^+$, $HE_k^- \cap OS^+$, $B^- \cap OS^+$ and $OS^- \cap B^+$. Unfortunately, in M^{12} all the examples constructed lie on submanifolds of high codimension, which does not permit one to draw conclusions regarding the measure of the corresponding subsets. Nevertheless, each of them contains a continuum (in the sense of power) of trajectories.

It is particularly interesting that total capture ($HE^- \cap B^+$) is possible. In contradistinction to the examples of Šmidt and Sitnikov, in this case a triple star is formed; on account of purely gravitational interaction, a double star is joined by a third body, flying in from infinity. This is analogous to the capture of a comet by the system Sun–Jupiter. We recall that in phase space the Lebesgue measure of a set filled by trajectories of motions with total capture is equal to zero (not knowing, apparently of Chazy's work, J. E. Littlewood independently proved an analogous assertion in [42]; see also [20]). Nonetheless, the study of this "sparse" set is interesting for at least two reasons. In the first place it is fairly complicated, and is distributed in a very intricate manner among the "rivulets" of the hyperbolic-elliptic motions. To understand the structure of the open set $HE^- \cap HE^+$ is scarcely possible without examining the structure of its boundary, an important part of which is formed by the set $HE^- \cap B^+$, and likewise $B^- \cap HE^+$ ("partial dispersal"), and also the oscillatory motions $HE^- \cap OS^+$ and $OS^- \cap HE^+$. If the conjecture formulated at the beginning of §4 is valid, then in the neighborhood of a typical point the set $HE^- \cap B^+$ must be in M^{12} the product of a set of measure zero (not closed, but probably F_σ) and an 11-dimensional disc. Second (see [16]), near a trajectory with complete capture there lie trajectories on which "temporary capture" takes place, and which may have practical interest.

For example, let the planet P_1 rotate about the star P_2 along an elliptic orbit. From infinity, along a nearly hyperbolic orbit, there comes flying a space ship P_3 (motion of the class HE_3^-). If the motion is sufficiently close to a motion of the class $\mathrm{HE}_3^- \cap \mathrm{B}^+$, then the space ship will remain in the system P_1P_2 for an arbitrarily long time, after which it will leave it and go off back to infinity. Poincaré in [8] predicted the possibility of such motions; a precise proof of their existence, let us say with an assigned number of approaches of the space ship to the planet, would be very interesting.

The structure of the set $\mathrm{B}^- \cap \mathrm{B}^+$ has been poorly explored, even though a significant portion of the publications on the three-body problem deal precisely with this. The theory of Kolmogorov, Arnol'd and Moser has made it possible to prove the existence of conditionally periodic motions in many nonintegrable problems of mechanics. In particular, in 1963 Arnol'd [18] showed that $\mathrm{B}^- \cap \mathrm{B}^+$ (for sufficiently small mass of two of the three bodies) contains a subset of positive measure, consisting of five-dimensional tori, filled with conditionally periodic motions (see also [6], [19] and [43]). According to [42] the sets $\mathrm{B}^+ \backslash \mathrm{B}^-$ and $\mathrm{B}^- \backslash \mathrm{B}^+$ have Lebesgue measure 0.

Analogous results are valid also for the general problem of the motion of a system of n bodies of planetary type. Let one of the bodies, which we call the "sun", have a mass much bigger than the other bodies ("planets"), which rotate about the sun in orbits which are approximately circular and which lie almost in one plane. If the masses of the planets are sufficiently small, and the eccentricities and inclinations of their orbits close to zero, then according to [18] the basic mass will consist of conditionally periodic motions. Although for the time being it is not possible to verify the "sufficient smallness" in application to the real solar system, so that Arnol'd's theorem cannot be applied here, all the same this theorem essentially complements the above-mentioned results of Lagrange and Laplace and makes the stability of the solar system, if not established, at least highly probable.

Conditionally periodic motions form the "regular" part of the sets $\mathrm{B}^- \cap \mathrm{B}^+$, but by no means exhaust it. In the set $(\mathrm{B}^- \cup \mathrm{OS}^-) \cap (\mathrm{B}^+ \cup \mathrm{OS}^+)$ there exists also a "quasirandom" part; we return to it later.

We recall that the following question remains open.

PROBLEM. Does OS form a set of positive Lebesgue measure in M^{12}? Does $\mathrm{OS}^\pm \cap \mathrm{HE}_k^\pm$? $\mathrm{OS}^+ \cap \mathrm{OS}^-$?

PART II
SYMBOLIC DYNAMICS

In recent years in the qualitative theory of differential equations considerable attention has been given to "irregularly constructed" objects, for whose description the classical analytic tools (asymptotic formulas, perturbation series, and so forth) turn out to be unsuitable. An example is the invariant set connected with "Smale's horseshoe", considered in A. B. Katok's lectures in this volume.

Here we had to do with a diffeomorphism of the two-dimensional sphere $f\colon S^2 \longrightarrow S^2$ which, hopefully, may be given by explicit formulas and may be of the class C^∞. A maximal set Λ, invariant under this diffeomorphism and contained in a square Q, an object quite naturally defined, is the product of two Cantor sets (and so is homeomorphic to a Cantor set), and the action of f on $\Lambda = \bigcap_{-\infty}^{+\infty} f^{-n}Q$ is topologically conjugate to the "baker's transformation". As we saw, this action is fairly complicated: in Λ there is an everywhere dense subset P consisting of periodic points of f; for any p_1 and p_2 in Λ there can be found a countable subset of points q such that the trajectory $\{f^n q\}$ is asymptotically equal as $n \longrightarrow +\infty$ to the trajectory $\{f^n p_1\}$, and as $n \longrightarrow -\infty$ to the trajectory $\{f^n p_2\}$, and so on. In addition, the set Λ is not decomposable, since there exist points $\lambda \in \Lambda$ whose trajectories are everywhere dense in Λ: $\overline{\{f^n(\lambda)\}} = \Lambda$.

It is hard to imagine how this complicated situation could be adequately described by analytical formulas in the classical spirit. At the same time, the action of the restriction $f_\Lambda = f|\Lambda$ on the set Λ is completely determined by the fact that f_Λ is topologically conjugate to the shift homeomorphism $T\colon \Omega^2 \longrightarrow \Omega^2$ in the space Ω^2 of doubly infinite binary sequences. The existence of a homeomorphism $\varphi\colon \Omega^2 \longrightarrow \Lambda$ with a commutative diagram

$$\begin{array}{ccc} \Omega^2 & \xrightarrow{\ T\ } & \Omega^2 \\ \varphi \big\downarrow & & \big\downarrow \varphi \\ \Lambda & \xrightarrow{\ f\ } & \Lambda \end{array} \qquad (1)$$

permits one to establish the properties just mentioned for the diffeomorphism f, as well as certain others. Thus the sequence $w = [\omega_n]$ of two symbols is a sort of code of the trajectory $\{f^n p\}$ of the point $p = \varphi(w)$; studying the codes and the shift, we study from a topological point of view the action of f on Λ.

Coding of the trajectories of smooth dynamical systems by sequences of natural numbers or sequences of symbols from some finite alphabet was apparently first applied to describing the global behavior of geodesics on surfaces of negative curvature (Hadamard, M. Morse and others; see, for example, [1], Chapter 8, §11). This served as an impetus to the study of the various properties of the shift homeomorphism in various subspaces of the space Ω^p of p-fold sequences. The whole circle of ideas and concepts connected with this received the name of "symbolic dynamics" [52]. However, some time afterwards the map $T\colon \Omega^p \longrightarrow \Omega^p$ was studied mainly from the point of view of ergodic theory, the more so since it is closely connected with ergodic dynamical systems of a probabilistic process—with Markov processes and, in particular, with the Bernoulli scheme. We return later to this connection.

On the other hand, in studying oscillations described by the second-order equation

$$\ddot{x} + f(x)\dot{x} + g(x) = p(t),$$

where f and g are suitable nonlinearities, and the perturbating force p is periodic, Levinson [53], Cartwright, and Littlewood [54] came across a situation in which the space Ω^2 also appeared naturally. In fact, in the examples of dissipative systems which they considered, described by the equation (2), the set of "limiting regimes" contains a subset which is in one-to-one correspondence with points from Ω^2 (an exposition of Levinson's work may be found in [55], §15).

Finally, in 1961 at the Kiev symposium on the theory of nonlinear oscillations, Smale [56] produced an example, an essential part of which was the celebrated "horseshoe".

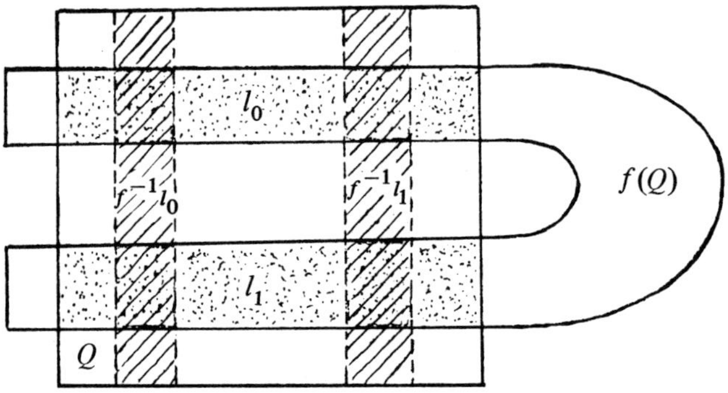

FIGURE 9

We recall that in "Smale's horseshoe" the intersection of a square Q with its image $f(Q)$ consists of two components l_0 and l_1 (Figure 9): $Q \cap f(Q) = l_0 \cup l_1$. Hence

$$\Lambda = \bigcap_{n=-\infty}^{+\infty} f^{-n}(Q) = \bigcap_{n=-\infty}^{+\infty} f^{-n}[Q \cap f(Q)] = \bigcap_{n=-\infty}^{+\infty} f^{-n}(l_0 \cup l_1)$$
$$= \bigcup_{[\omega_n]} \bigcap_{n=-\infty}^{+\infty} f^{-n} l_{\omega_n},$$

where $[\omega_n]$ is all possible sequences of zeros and ones. It is essential that for each sequence $w = [\omega_n] \in \Omega^2$ the intersection $\bigcap_{-\infty}^{\infty} f^{-n}\, l_{\omega_n}$ consist of just one point and that the mapping $w \mapsto \varphi(w) = p = \bigcap_{-\infty}^{\infty} f^{-n}\, l_{\omega_n}$ be a homeomorphism of Ω^2 onto Λ, equivariant in the sense that $\varphi \circ T = f \circ \varphi$, i.e. that the diagram (1) be commutative.

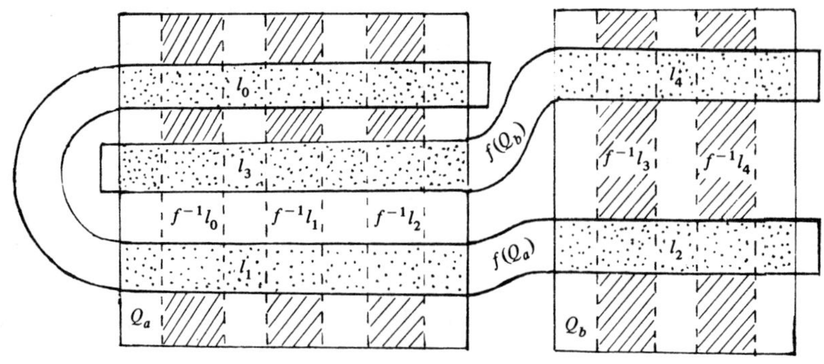

FIGURE 10

Naturally, one may modify this construction by various devices. Let us consider, for example, the situation depicted in Figure 10. The set Q now consists of two rectangles Q_a and Q_b, and the intersection $Q \cap f(Q)$ of five components

$$Q \cap f(Q) = \bigcup_{i=0}^{4} l_i.$$

As in the case of the "horseshoe", it is assumed that Q_a, Q_b, l_i and $f^{-1}l_i$ are rectangles, which in the sequel are to be conceived as lying in a plane with their sides parallel to the coordinate axes. The adjectives "horizontal" and "vertical" relate precisely to this situation. The mappings $f\colon f^{-1}l_i \longrightarrow l_i$ are assumed linear (hyperbolic rotations), compressing in the vertical and stretching in the horizontal directions. Once more, let Λ be a maximal invariant set contained in Q. Then, as above,

$$\Lambda = \bigcap_{-\infty}^{+\infty} f^{-n}(Q) = \bigcup \bigcap_{-\infty}^{+\infty} f^{-n} l_{\omega_n},$$

but here the symbols ω_n belong to an alphabet of the letters 0, 1, 2, 3, 4. Each point $p \in \Lambda$ may be put into correspondence with precisely one sequence $w = [\omega_n]$ in such a way that $p \in \bigcap_{-\infty}^{\infty} f^{-n} l_{\omega_n}$. However, now it is no longer the case that for every sequence $w \in \Omega^5$ the intersection $\bigcap_{-\infty}^{\infty} f^{-n} l_{\omega_n} \neq \varnothing$. To see this, we consider the intersection

$$f^{-1}(Q) \cap Q \cap f(Q) = f^{-1}(Q \cap f(Q)) \cap Q \cap f(Q)$$

$$= f^{-1}\Big(\bigcup_{i=0}^{4} l_i\Big) \cap \bigcup_{j=0}^{4} l_j = \bigcup_{i,j} f^{-1} l_i \cap l_j.$$

The sets $f^{-1}l_0$, $f^{-1}l_1$ and $f^{-1}l_2$ are contained in Q_a and so do not intersect with l_2 or l_4, which lie in Q_b. On the other hand, $f^{-1}l_3$ and $f^{-1}l_4$ lie in Q_b and intersect with l_2 and l_4, but not with l_0, l_1 and l_3.

Hence in the expression $\bigcup_{i,j} f^{-1}l_i \cap l_j$ the intersections $f^{-1}l_i \cap l_j$ will be nonempty not for all pairs (i, j), but only for those for which in the matrix

$$\Pi = (\pi_{ij},\ 0 \leqslant i, j \leqslant 4) = \begin{pmatrix} 1 & 1 & 0 & 1 & 0 \\ 1 & 1 & 0 & 1 & 0 \\ 1 & 1 & 0 & 1 & 0 \\ 0 & 0 & 1 & 0 & 1 \\ 0 & 0 & 1 & 0 & 1 \end{pmatrix} \qquad (3)$$

the corresponding elements π_{ij} are equal to 1. In the same way one shows that if $\bigcap_{-\infty}^{\infty} f^{-1} l_{\omega_n} \neq \varnothing$, then $\pi_{\omega_n \omega_{n+1}} = -1$ for all n.

We denote by Ω^Π the subset in Ω^5 consisting of all sequences $w = [\omega_n]$ for which $\pi_{\omega_n \omega_{n+1}} = 1$ for all n. The space Ω^5 may be equipped with the product topology; in this topology a neighborhood of the sequence $w^{(0)}$ consists of all sequences w coinciding with $w^{(0)}$ on some (fixed for the given topology) range of numbers $n_1 \leqslant n \leqslant n_2$.

It is easy to see that

$$w^{(0)} = \lim_{k\to\infty} w^{(k)},$$

where $w^{(0)} = [\omega_n^0]$ and $w^k = [\omega_n^k]$, if and only if, for each n, starting at some point (different for different n) we have $\omega_n^k = \omega_n^0$. In this topology the space Ω^5 is compact, and the subset Ω^Π is closed, and so also compact; $T\colon \Omega^5 \longrightarrow \Omega^5$ is a homeomorphism with respect to which Ω^Π is invariant.

In analogy with the theory of probability, the homeomorphism $T\colon \Omega^\Pi \longrightarrow \Omega^\Pi$ is called a *topological Markov chain* (TMC), and the matrix $\Pi = (\pi_{ij})$ the *matrix of admissible transitions* of this chain.

Arguing as in the case of the "horseshoe", we show that the mapping $w = [\omega_n] \mapsto \bigcap_{-\infty}^{+\infty} l_{\omega_n} = p = \varphi(w)$ is a homeomorphism between Ω^Π and $\Lambda = \bigcap_{-\infty}^{+\infty} f^{-n}Q$. Here

$$(f\circ\varphi)(\omega) = f\left(\bigcap_{-\infty}^{+\infty} f^{-n} l_{\omega_n}\right) = \bigcap_{-\infty}^{+\infty} f^{-(n-1)} l_{\omega_n}$$

$$= \bigcap_{-\infty}^{+\infty} f^{-n} l_{\omega_{n+1}} = \varphi([\omega_{n+1}]) = (\varphi\circ T)(\omega),$$

where the homeomorphism φ is equivariant and the diagram

$$\begin{array}{ccc} \Omega^\Pi & \xrightarrow{\;T\;} & \Omega^\Pi \\ \varphi\big\downarrow & & \big\downarrow\varphi \\ \Lambda & \xrightarrow{\;f\;} & \Lambda \end{array}$$

is commutative.

With the one-sided sequences $w^+ = [\omega_n\colon n \geqslant 0]$ and $w^- = [\omega_n\colon n \leqslant -1]$ we can associate the intersections

$$W^s_{\mathrm{loc}}(p) = \bigcap_{n\geqslant 0} f^{-n} l_{\omega_n}, \quad W^u_{\mathrm{loc}}(p) = \bigcap_{n\leqslant -1} f^{-n} l_{\omega_n}.$$

The first of them is a vertical, and the second a horizontal segment, passing in Q through the point

$$p = \varphi(w) = \bigcap_{-\infty}^{+\infty} f^{-n} l_{\omega_n},$$

corresponding to the whole sequence $w = w^- \times w^+$, compounded from w^- and w^+. It is evident from the definition that W^s_{loc} and W^u_{loc} satisfy the relations

$$f[W^s_{\mathrm{loc}}(p)] \subseteq W^s_{\mathrm{loc}}[f(p)], \qquad f(W^u_{\mathrm{loc}}(p)) \supseteq W^u_{\mathrm{loc}}(f(p)). \tag{4}$$

The tangents $TW^s_{\mathrm{loc}}(p)$ and $TW^u_{\mathrm{loc}}(p)$, drawn at every point $p \in \Lambda$, determine on Λ (as in the case of a "horseshoe") the structure of a hyperbolic set (see Katok's lectures contained in this volume). W^s_{loc} and W^u_{loc} themselves are local sections of the sheets of invariant layers connected with the hyperbolic set Λ.

The local diffeomorphism $f\colon Q \longrightarrow \mathbf{R}^2$ is easily extended to a global diffeomorphism of the two-dimensional sphere. The first stage in the corresponding construction —the mapping of a disc into itself—is represented in Figure 11. One may continue the

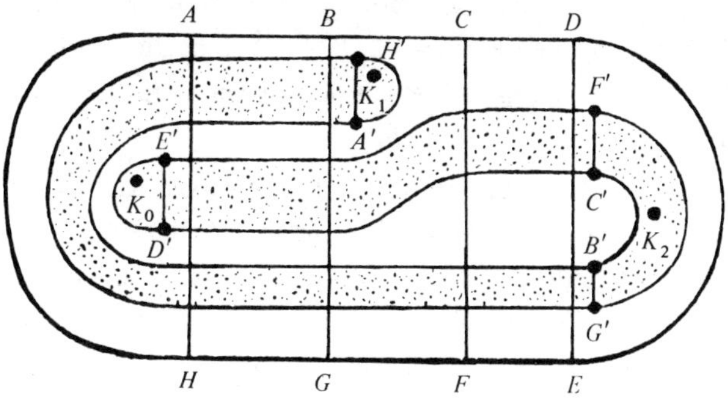

FIGURE 11

process as in the case of the horseshoe, extending f so that outside the disc it has a unique point of repulsion q.

It is easy to see that the set of nonwandering points of the diffeomorphism $f\colon S^2 \longrightarrow S^2$ thus formed consists of the point of repulsion q of the hyperbolic set Λ, and an attracting trajectory of period 3 (the points k_0, k_1 and k_2 on Figure 11). This diffeomorphism may be taken to be infinitely smooth, and for it the conditions of Robbin's theorem (see Katok's lectures) will hold. Hence f is C^2-rough, i.e. a whole neighborhood of it in the C^2-topology consists of diffeomorphisms which are topologically conjugate to f.

In order to see more graphically which sequences $w = [\omega_n] \in \Omega^5$ are "admissible", i.e. belong to Ω^n, it is convenient to use the "flow chart" (Figure 12). In this

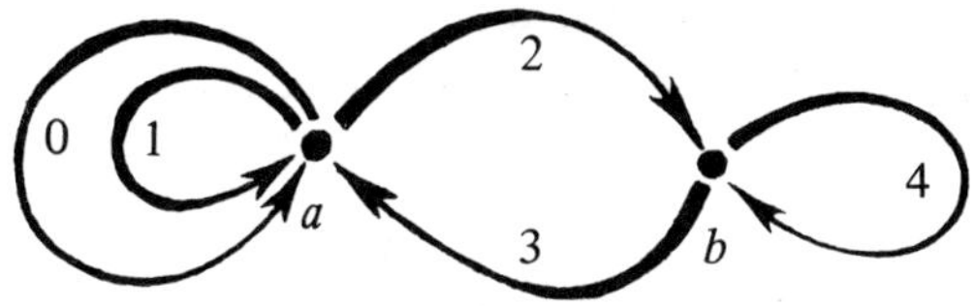

FIGURE 12

graph the vertices a and b symbolize the squares Q_a and Q_b of Figure 10, and the edges, numbered from 0 to 4, symbolize the intersections l_i, numbered respectively. It is not hard to see that $f^{-1}l_i \cap l_j$ is nonempty ($\pi_{ij} = 1$) if and only if in the graph of Figure 12 the end of an edge with index i is the beginning of an edge with index j. Hence it follows that $p = \bigcap_{-\infty}^{+\infty} f^{-n} l_{\omega n} \in \Lambda$ if and only if the sequence $[\omega_n]$ is the sequence of indices of edges of some two-sidedly infinite path in the graph of Figure 12. In just the same way $W^s_{\mathrm{loc}}(p)$ corresponds to a path infinite in the forward sense, with initial edge ω_0, and $W^u_{\mathrm{loc}}(p)$ corresponds to a path infinite in the backward sense, with last edge ω_{-1}.

We now show that the homeomorphism $T\colon \Omega^{\Pi} \longrightarrow \Omega^{\Pi}$ is topologically transitive and that its periodic points are dense in Ω^{Π}. Since the restriction of f to Λ is topologically conjugate to $T \mid \Omega^{\Pi}$, f on Λ will have the same properties.

Let $w^{(0)} = [\omega_n^0]$ be an arbitrary sequence in Ω^{Π} and U some neighborhood of it, consisting of all possible sequences $w = [\omega_n] \in \Omega^{\Pi}$ coinciding with $w^{(0)}$ on a segment of indices $n_1 \leqslant n \leqslant n_2$. Let this segment in $w^{(0)}$ begin with the symbol x and end with the symbol y.

Let us consider the graph in Figure 12. It is obvious that for any pair l', l'' of its edges the graph contains a path in which l' is the initial edge and l'' the final one. In particular, there exists a path γ consisting of k edges, the initial edge of which has the index y and the last the index x. We now form a periodic sequence $w^{(1)} = [\omega_n^{(1)}] \in U \cap \Omega^{n}$ in the following manner. On the segment $n_1 \leqslant n \leqslant n_2$ we set $\omega_n^{(1)} = \omega_n^{(0)}$ (otherwise $w^{(1)}$ does not fall in U). The symbols $\omega_{n_2}^{(1)} = y$, $\omega_{n_2+1}^{(1)}, \ldots, \omega_{n_2+k-1}^{(1)} = x$ we take equal to the indices of the edges appearing in the path γ. This block of $n_2 - n_1 + k$ symbols corresponds to a closed path in the graph. The whole sequence $w^{(1)}$ is now obtained by periodic repetition of this block.

Thus any neighborhood of any point $w^{(0)} \in \Omega^{\Pi}$ contains a periodic sequence from Ω^{Π}, and it is therewith proved that the periodic points of the homeomorphism T are dense in Ω^{Π}. Using similar considerations, one may construct a sequence $\hat{w} = [\hat{\omega}_n]$ whose trajectory (and indeed only the half-trajectory $\{T^n \hat{w}, n \geqslant 0\}$) is everywhere dense in Ω^{Π}.

We number in arbitrary order all possible finite blocks $A_1, A_2, \ldots$, in which symbols from 0 to 4 are in an admissible order (i.e. for neighboring symbols i and j, $\pi_{ij} = 1$, or the sequence of edges in the graph in Figure 12 corresponding to the block forms a path). Let the block A_k begin with the symbol a_k and end with the symbol b_k. As already indicated, there exists an admissible block of symbols B_k beginning with b_k and ending with a_{k+1}. We now take the sequence

$$\hat{w} = \underset{n<0}{\cdots\cdots}\, \overbrace{a_1 \cdots b_1}^{A_1} \underbrace{\cdots}_{B_1} \overbrace{a_2 \cdots b_2}^{A_2} \underbrace{\cdots a_3}_{B_2} \cdots\cdots,$$

in which the elements $\hat{\omega}_n$, $n < 0$, are arbitrary, $\omega_0 = a_1$, and there follow in turn the blocks $A_1, B_1, A_2, B_2, \ldots$, so that the end of one block is the beginning of the next.

Now let U be a neighborhood of an arbitrary point $w^{(0)} \in \Omega^{\Pi}$, consisting of sequences which coincide with $w^{(0)}$ on a segment of indices $n_1 \leqslant n \leqslant n_2$. Let the block $\omega_{n_1}^0, \omega_{n_1+1}^0, \ldots, \omega_{n_2}^0$, which we have numbered A_k, and its beginning a_k stand in $\hat{w}$ in the location with index N. Then $T^N \hat{w} \in U$. Hence the positive half-trajectory $\{T^n \hat{w}, n \geqslant 0\}$ of the point $\hat{w}$ will sooner or later fall in any neighborhood of any point $w^{(0)}$, i.e. will be everywhere dense. Hence $T \mid \Omega^{\Pi}$ is topologically transitive, as asserted.

The diffeomorphism $f\colon S^2 \to S^2$ just considered may appear too artificial. However, symbolic dynamics does admit more natural examples.

We now take an automorphism $f\colon \mathbf{T}^2 \to \mathbf{T}^2$ of the torus, appearing in the lectures of Kušnirenko and Katok, given by the formulas

$$\left.\begin{aligned} x' &= 2x + y \\ y' &= x + y \end{aligned}\right\} \pmod 1. \tag{5}$$

As a linear map of the plane, (5) is a hyperbolic rotation with the fixed point $A(0, 0)$ and the invariant lines

$$\text{a magnifying separatrix} \quad y = kx, \tag{6}$$

$$\text{a contracting separatrix} \quad y = -x/k \tag{7}$$

(here $k = \tfrac{1}{2}(\sqrt{5} - 1) = 0.618\ldots$, the golden section).

We consider in the plane $X0Y$ the rectangle $Q_a = ABCD$, whose sides AB and AD are segments of the separatrices (6) and (7), while BC and CD are segments of lines parallel to these separatrices passing through $(1, 0)$ and $(1, 1)$ (we remark that all vertices of the unit square constitute one and the same point on the torus, and so AD and BC are segments of one and the same separatrix; the same applies to AB and CD), and $Q_b = DEFG$, whose side DG lies on AD, EF and ED being segments of lines parallel to the separatrices (6) and (7) and passing through $(1, 1)$, while FG lies on a line parallel to the separatrix (7) and passing through $(0, 1)$ (again all these are various segments of one and the same pair of separatrices on the torus).

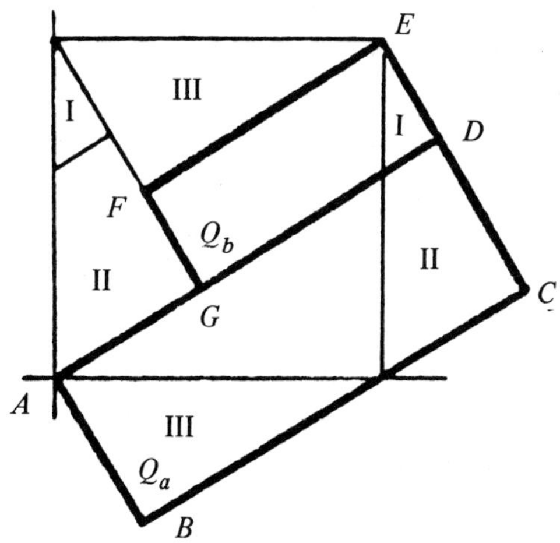

FIGURE 13

It is easy to see (Figure 13) that $Q_a \cup Q_b$ covers the whole torus (for clarity, parts of the plane forming one and the same subset of the torus have been marked with the same Roman numerals), while Q_a and Q_b do not have common interior points.

We now consider what the automorphism (5) does with Q_a and Q_b. On the plane it contracts in the direction of the separatrix (7) and magnifies in the direction of the separatrix (6) by a factor $\lambda = k + 2 = 2.618\ldots$ (see Figure 14). The images of the points $B, C, \ldots$ are indicated by primes ($A' = A$ and $G' = D$). The images $f(Q_a)$ and $f(Q_b)$ once more cover the whole torus, and the intersection of Q_a and Q_b with their images may be divided into five rectangles l_i, $0 \leqslant i \leqslant 4$.

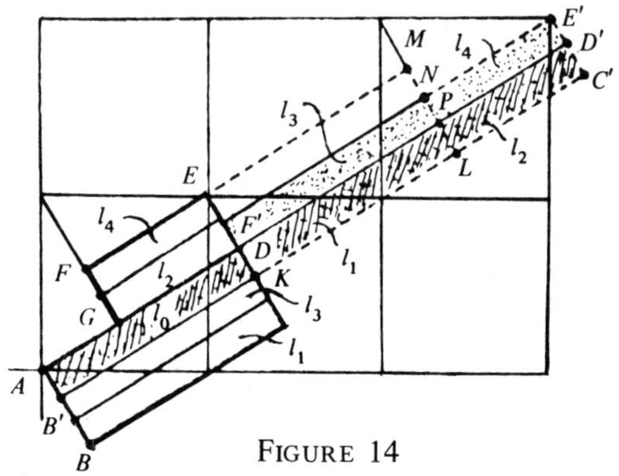

FIGURE 14

In fact, $f(Q_a)$ is formed on the plane by the rectangle $AB'C'D' = l_0 \cup l_1 \cup l_2$, where $l_0 = AB'KD \subset f(Q_a) \cap Q_a$; $l_1 = KDPL$ is obtained in the plane as the intersection of $f(Q_a)$ with the rectangle $EKLM$, and on the torus is again a part of $f(Q_a) \cap Q_a$ (since $EKLM$ is obtained from $ABCD$ by a parallel integer-valued shift, and on the torus this is one and the same set); and $l_2 = LPD'C' \subset f(Q_a) \cap Q_b$.

In just the same way, $f(Q_b) = F'DD'E' = l_3 \cup l_4$, where $l_3 \subset f(Q_b) \cap Q_a$ and $l_4 = f(Q_b) \cap Q_b$. We see now that the intersections of $f(Q_a)$ and $f(Q_b)$ with Q_a and Q_b are formed just as in Figure 10; however, now $f(Q_a \cup Q_b) = Q_a \cup Q_b = \mathbf{T}^2$, so that no "superfluous" points remain; in return, certain points of the torus will be covered twice (on Figure 10 the various l_i do not intersect pairwise, and now they can intersect, though only on the boundary). The maps $f\colon f^{-1}l_i \to l_i$ are hyperbolic rotations, and we may carry out the same constructions as were applied to the investigation of "Smale's horseshoe" and the diffeomorphism of Figures 10 and 11. Again the space Ω^Π appears, where Π is the matrix (3), and the TMC (topological Markov chain) $T\colon \Omega^\Pi \to \Omega^\Pi$.

THEOREM [57]. *There exists a map $\varphi\colon \Omega^\Pi \to \mathbf{T}^2$, which is continuous and homeomorphic on the completion of a set of Baire first category and such that $f \circ \varphi = \varphi \circ T$ and the diagram*

$$\begin{array}{ccc} \Omega^\Pi & \xrightarrow{\ T\ } & \Omega^\Pi \\ {\scriptstyle\varphi}\downarrow & & \downarrow{\scriptstyle\varphi} \\ T^2 & \xrightarrow{\ f\ } & T^2 \end{array} \tag{8}$$

is commutative.

The above-mentioned set of first category is obtained as the preimage of the sides of the rectangles Q_a and Q_b and all of their translates under the automorphism f. The preimage of each side is nowhere dense in Ω^{Π}, and so the set where φ is not a homeomorphism is a countable union of nowhere dense sets, which is indeed the definition of a "set of first category".

What sort of conclusions can we draw from this theorem? First of all, we saw already that the periodic points of the shift T are dense in Ω^{Π}. If $T^n w = w$, then it is clear from the diagram (8) that $f^n[\varphi(w)] = \varphi(T^n w) = \varphi(w)$, and so $p = \varphi(w)$ is a periodic point of the automorphism f. Let q be any point of the torus and $w^{(0)} \in \varphi^{-1}(q)$. By what was proved above, there exists a sequence of periodic points $w^{(k)} \longrightarrow w^{(0)}$; but then $\varphi(w^{(k)})$ is a sequence of periodic points of the automorphism f, which converges to $\varphi(w^{(0)}) = q$. Hence the periodic points of f are dense in $\mathbf{T}^2$.

In just the same way one proves that $\mathbf{T}^2$ contains an everywhere dense (half) trajectory, in virtue of which f is topologically transitive. Having analyzed which points in Ω^{Π} may be identified under the mapping φ, we can express any topological question relating to the properties of f in the terminology of "topological dynamics". In many cases this approach turns out to be very suitable.

Furthermore, in Ω^{Π} we consider all possible "cylinders"

$$C_{a_m a_{m+1} \cdots a_{m+k}} = \{w = [\omega_n]\,;\, \omega_m = a_m, \ldots, \omega_{m+k} = a_{m+k}\}. \tag{9}$$

Under φ such a cylinder goes into the intersection

$$f^{-m} l_{a_m} \cap f^{-(m+1)} l_{a_{m+1}} \cap \cdots \cap f^{-(m+k)} l_{a_{m+k}}.$$

Let μ denote the usual Lebesgue measure (area), transferred from the plane to the torus. We attach to each cylinder (9) a measure, denoted by ν:

$$\nu(C_{a_m \cdots a_{m+k}}) = \mu\left(\bigcap_{n=m}^{m+k} f^{-n} l_{a_n}\right). \tag{10}$$

It is not hard to verify that the measure thus defined on all cylinders satisfies the conditions of the well-known theorem of Kolmogorov, so that ν can be extended to all Borel sets in Ω^{Π}. The automorphism (5) preserves area: $\mu(fA) = \mu A$, since the determinant of its matrix $\det\binom{2\ 1}{1\ 1} = 1$; the mapping T has the same property. To prove this we take again an arbitrary cylinder $C_{a_m a_{m+1} \cdots a_{m+k}}$, whose image will be the cylinder

$$T(C_{a_m \cdots a_{m+k}}) = C_{a'_{m+1} \cdots a'_{m+k+1}},$$

where $a'_i = a_{i-1}$. It follows from (10) that

$$\nu(TC_{a_m \cdots a_{m+k}}) = \mu\left(\bigcap_{n=m+1}^{m+k+1} f^{-n} l_{a_{n-1}}\right)$$

$$= \mu\left(f \bigcap_{n=m}^{m+k} f^{-n} l_{a_n}\right) = \mu\left(\bigcap_{n=m}^{m+k} f^{-n} l_{a_n}\right).$$

In Ω^{Π} we define a new measure $\widetilde{\nu}$ by the equation $\widetilde{\nu}(A) = \nu(TA)$. Then on all cylinders we have

$$\widetilde{\nu}(C_{a_m \cdots a_{m+k}}) = \nu(TC_{a_m \cdots a_{m+k}}) = \nu(C_{a_m \cdots a_{m+k}}).$$

It follows from Kolmogorov's theorem that in this case $\widetilde{\nu} = \nu$, i.e. $\nu(TA) = \nu(A)$, which means that the measure ν is invariant.

One may deduce from (10) that for any Borel subset of the torus we have $\nu(\varphi^{-1}(B)) = \mu B$. Sets in Ω^{Π} and $\mathbf{T}^2$ on which φ is not one-to-one have measure 0. Since in the metric theory of dynamical systems ("ergodic theory") it is customary to neglect sets of measure 0, from this point of view φ is an isomorphism of the measure spaces (Ω^{Π}, T, ν) and $(\mathbf{T}^2, f, \mu)$. Here the measure ν in Ω^{Π} is none other than the probability distribution of a certain Markov chain. This means that one can find q_i and p_{ij}, $0 \leqslant i, j \leqslant 4$, such that for any cylinder

$$\nu(C_{a_0 \cdots a_n}) = q_{a_0} p_{a_0 a_1} p_{a_1 a_2} \cdots p_{a_{n-1} a_n}.$$

The matrix (p_{ij}) is called in probability theory a *matrix of transition probabilities*, and q_i the *stationary state probabilities* of the given chain.

By analogous arguments, Adler and Weiss [57] showed that any ergodic automorphism of a two-dimensional torus given by an integral unimodular matrix is isomorphic, from the point of view of ergodic theory, to some stationary Markov chain. This isomorphism enables them to carry out a classification of all such automorphisms. Here the unique invariant turned out to be the entropy of an automorphism in the sense of Kolmogorov, equal to the logarithm of the modulus of that eigenvalue of the matrix of the automorphism for which this modulus exceeds 1 (for an ergodic automorphism of the torus one of the eigenvalues is necessarily such).

One further situation in which the methods of symbolic dynamics can be applied is connected with so-called homoclinic points. Let p be a hyperbolic fixed point of the diffeomorphism $f\colon M \longrightarrow M$ of an arbitrary manifold, and let $W^s(p)$ and $W^u(p)$ be its invariant stable and unstable manifolds (see Kušnirenko's lectures). Then $q \in W^s(p) \cap W^u(p)$ is called a *homoclinic point* of the point p. Its trajectory $\{f^n q\}$ is doubly asymptotic in the sense that $f^n(q) \longrightarrow p$ both as $n \longrightarrow +\infty$ and as $n \longrightarrow -\infty$. If the manifolds $W^s(p)$ and $W^u(p)$ intersect at the point q transversally (i.e. the tangent planes $T_q W^s(p)$ and $T_q W^u(p)$ generate the whole tangent space $T_q M$ of the point q), then q is called a *transversal homoclinic point.* In Kušnirenko's lectures it has already been noted that homoclinic points are connected with an extremely involved picture of the type illustrated in Figure 15. Symbolic dynamics makes it possible to represent the behavior of f near the orbit of a transversal homoclinic point in a rather revealing form.

We consider the compact set Γ consisting of trajectories of the points q and p, and cover it with a finite number of sufficiently small neighborhoods in the following manner (Figure 16). First p is covered by some neighborhood U_0. Apart from p, this

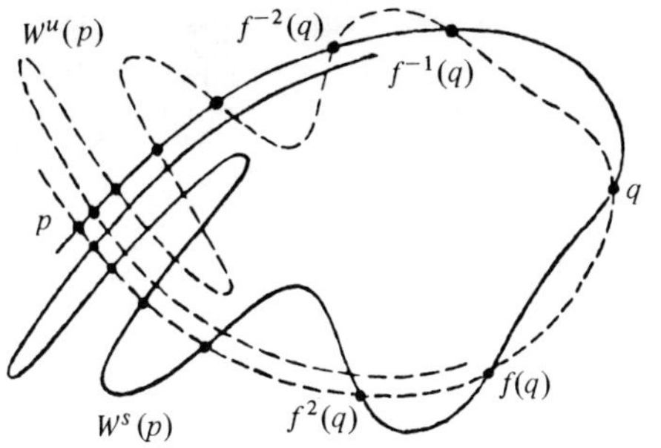

FIGURE 15

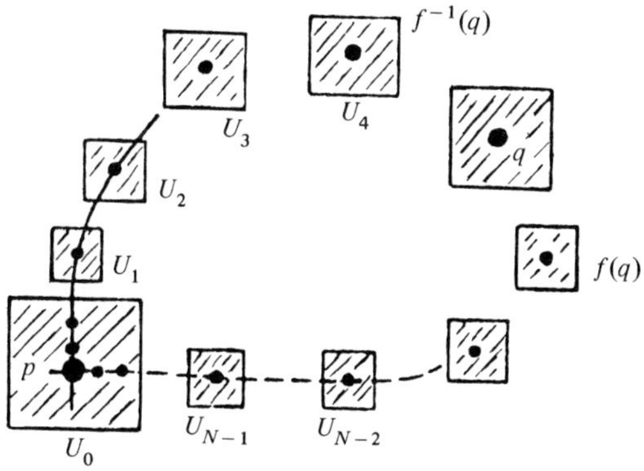

FIGURE 16

neighborhood will cover all but a finite number of points of the trajectory $\{f^n(q)\}$. These remaining points are each covered by pairwise nonintersecting neighborhoods $U_1, \ldots, U_{N-1}$.

From this construction it follows at once that the intersections

$$f(U_0) \cap U_0, f(U_0) \cap U_1, f(U_1) \cap U_2, \ldots, f(U_{N-2}) \cap U_{N-1}, f(U_{N-1}) \cap U_0 \quad (11)$$

are nonempty (the first intersection contains p, and the others contain at least one point of the trajectory $\{f^n q\}$). We can depict this situation in the form of a graph (Figure 17), the edges $\overline{ij}$ of which correspond to intersections $f(U_i) \cap U_j$, and also with the aid of the matrix

$$\Pi(N) = \begin{pmatrix} 1 & 1 & 0 & & & & & \\ 0 & 0 & 1 & 0 & & & & \\ 0 & 0 & 0 & 1 & 0 & & & \\ \cdot & \cdot & \cdot & \cdot & \cdot & \cdot & \cdot & \\ & & & & & & \cdot & 0 \\ 0 & 0 & 0 & \cdot & \cdot & \cdot & 0 & 1 \\ 1 & 0 & 0 & \cdot & \cdot & \cdot & 0 & 0 \end{pmatrix}. \quad (12)$$

As before, to this matrix there corresponds a subset $\Omega(N)$ in the space of sequences Ω^N of symbols $0, 1, \ldots, N-1$; $\Omega(N)$ contains those sequences $[\omega_n]$ for

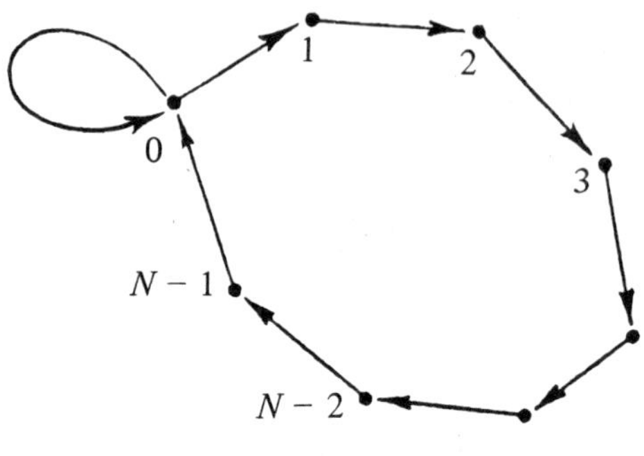

FIGURE 17

which in consecutive places there stand only those pairs of symbols $\omega_n\omega_{n+1}$ which correspond to an edge in the graph of Figure 17 or to a unity in the matrix (12). (Observe that the graph of Figure 17 is used somewhat differently from that in Figure 12.) The subset $\Omega(N)$ is invariant with respect to the shift homeomorphism T; this gives one a new TMC $T\colon \Omega(N) \longrightarrow \Omega(N)$.

THEOREM [58]. *For every open set V containing the set Γ, one may choose neighborhoods $U_0, \ldots, U_{N-1}$ in such a way that*:

1) $U = U_0 \cup U_1 \cup \cdots \cup U_{N-1} \subset V$;

2) *the maximal invariant set* $\Lambda = \bigcap_{-\infty}^{\infty} f^{-n}U$ *contained in U is a hyperbolic set for f; and*

3) *the restriction $f \mid \Lambda$ is topologically conjugate to the TMC $T\colon \Omega(N) \longrightarrow \Omega(N)$, i.e. there exists a homeomorphism $\varphi\colon \Omega(N) \longrightarrow \Lambda$ such that the diagram*

$$\begin{array}{ccc} \Omega(N) & \xrightarrow{T} & \Omega(N) \\ {\scriptstyle\varphi}\downarrow & & \downarrow{\scriptstyle\varphi} \\ \Lambda & \xrightarrow{f} & \Lambda \end{array} \tag{13}$$

is commutative.

The number N depends on the choice of the neighborhood V and increases as the latter decreases. The graph of Figure 17 and the matrix (12) represent schematically the intersections (11). In general, other intersections $f(U_i) \cap U_j$ may also be nonempty, and one could ask about extending the space of admissible sequences $\Omega(N)$, with corresponding additions of new edges to the graph in Figure 17 (or likewise changing some zeros to ones in the matrix (12)).

Such an extension is, however, not always productive. The basic stage in the construction of the homeomorphism $\varphi\colon \Omega(N) \longrightarrow \Lambda$ is linked to the proof that

$$\bigcap_{-\infty}^{+\infty} f^{-n}U_{\omega_n} \neq \varnothing$$

when all pairwise intersections $fU_{\omega_n} \cap U_{\omega_{n+1}}$ are admissible (the "Markov property"). For "Smale's horseshoe" and the diffeomorphism in Figure 10 this assertion followed easily from the linearity of the mapping $f: f^{-1}l_i \longrightarrow l_i$ and simple geometric considerations. In the general situation, which we encounter in the neighborhood of the trajectory of a homoclinic point, the proof of the analogous "Markov property" becomes nontrivial. For its validity we need that the intersections $fU_i \cap U_j$ should be not only nonempty, but also sufficiently "good". Without going into details, we illustrate this in Figures 18 and 19; for more detail on what is meant by a "good" intersection see [59].

On the other hand, the above theorem asserts that Λ is a maximal invariant set in some neighborhood of it, and so it can be extended only at the expense of additional trajectories that diverge far enough from Λ. The matter was in no way the same in the case of the original set Γ. It is also invariant, and from the transversality of the intersection $W^s(p) \cap W^u(p)$ at a homoclinic point one may deduce that Γ is a hyperbolic set. However, it is not "locally maximal", since the theorem at the same time asserts that any neighborhood V containing Γ contains a larger invariant set $\Lambda \supset \Gamma$.

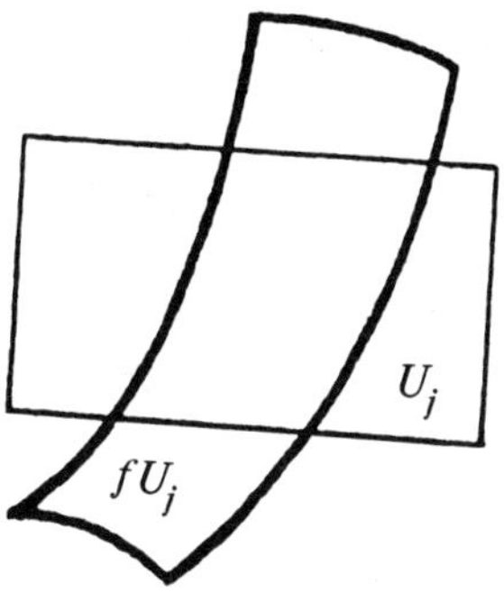

"Good" intersection

FIGURE 18

We pass to a discussion of the consequences which may be inferred from the above theorem. As in previous cases and with the same arguments, it may be proved that f acts on Λ in a topologically transitive manner, and that the periodic points are dense in Λ. Every trajectory $f^n x$, $x \in \Lambda$, can be coded in a one-to-one manner by a sequence $w = [\omega_n] \in \Omega(N)$. With the aid of Figure 17 it is easy to see that that this sequence is constructed as follows. There will always be zeros in it; after each zero can stand either a zero or a unity, where the latter will always be the beginning of a block $\sigma = 1\ 2\ \ldots\ (N-1)0$. Thus w consists of zeros, interspersed with blocks σ (not excluding that several or even an infinite number of blocks σ can come in succession):

$$w = \cdots 0\overset{\overset{\omega_0}{\downarrow}}{0}0\ \overbrace{12\cdots(N-1)0}^{\sigma}\ \overbrace{12\cdots(N-1)0}^{\sigma}\ 000\ \overbrace{12\cdots(N-1)0}^{\sigma}\cdots. \qquad (14)$$

The number of blocks σ may be finite or infinite; they can appear in the sequence w infinitely far to the right, or to the left, or in both directions.

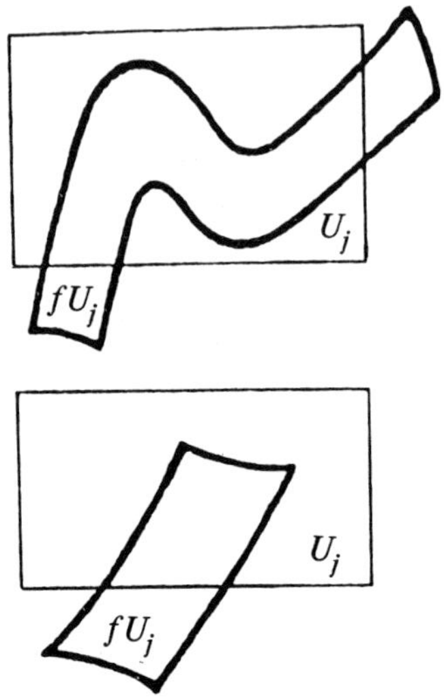

"Bad" intersections

FIGURE 19

Under the mapping φ the sequence w goes into the point $\varphi(w) \in U_{\omega_0}$. As an example, we construct w as in (14). To it there corresponds a point $x \in U_0$. By (13)

$$f(x) = f(\varphi(w)) = \varphi(Tw) \in U_{(Tw)_0} = U_{\omega_1} = U_0.$$

Similarly, $f^2(x) \in U_1$, and subsequent iterations $f^n(x)$ traverse twice in succession all the neighborhoods $U_2, \ldots, U_{N-1}, U_0, U_1, \ldots$; then $f^n(x)$ will fall three times in U_0, and so on.

Let $s \geqslant 2$ and $k = N + s - 2$. We consider the subset $\Omega^{(s)} \subseteq \Omega^{(N)}$, invariant with respect to T^k and consisting of sequences for which $\omega_0 = 0$. By virtue of the invariance, in all places whose index is divisible by k there must be a zero: $\omega_{mk} = 0$, and to describe a $w \in \Omega^{(s)}$ it is sufficient to clarify what blocks of $k - 1$ symbols can appear between two zeros. After turning once more to Figure 17, we find that only the following s blocks can have this property:

$$A_0 = 0 \;\vdots\; \overbrace{0 \cdots 0}^{s-1} \;\vdots\; 0$$

$$A_1 = 0 \;\vdots\; 12 \cdots (N-1)\, \overbrace{00 \cdots 0}^{s-2} \;\vdots\; 0$$

$$A_2 = 0 \;\vdots\; 012 \cdots (N-1)\, \overbrace{00 \cdots 0}^{s-3} \;\vdots\; 0$$

$$\cdots\cdots\cdots\cdots\cdots\cdots\cdots\cdots$$

$$A_{s-1} = 0 \;\vdots\; \underbrace{00 \cdots 0}_{s-2}\, 12 \cdots (N-1) \;\vdots\; 0$$

Thus, each sequence $w \in \Omega^{(s)}$ has the form

$$\ldots 0A_{i_{-1}}0A_{i_0}0A_{i_1}0A_{i_2}0 \ldots, \tag{15}$$

where the indices i_n can take independently any value $0, 1, \ldots, s-1$. It is clear that $\Omega^{(s)}$ is in one-to-one correspondence with the set $\Omega^{(s)}$ of all sequences of s symbols. The mapping T^k shifts (15) s symbols to the left, so that the block $A_{i_{n+1}}$ appears in the place of A_{i_n}. This is precisely what the shift T does to $\Omega^{(s)}$. This shows that the homeomorphism $T^k\colon \Omega^{(s)} \longrightarrow \Omega^{(s)}$ is conjugate to $T\colon \Omega^{(s)} \longrightarrow \Omega^{(s)}$. The mapping $\varphi\colon \Omega^{(s)} \longrightarrow \Lambda$ transports this conjugacy to the manifold. Since for all sequences in $\Omega^{(s)}$ we have $\omega_0 = 0$, it follows that $\varphi(\Omega^{(s)}) \subset U_0$, and we obtain the

COROLLARY. *If p is a hyperbolic fixed point of the diffeomorphism $f\colon M \longrightarrow M$, having a transversal homoclinic point, then for every neighborhood $U(p)$ and every integral s there exist a k and a subset $\Lambda^s \subset U(p)$, invariant with respect to f^k, such that the restriction $f^k \mid \Lambda^s$ is topologically conjugate to the shift homeomorphism in the space of s-fold sequences. (For $s = 1$ this assertion is trivial.)*

This is how Smales's original result [60] was formulated, showing the possibility of using the methods of "symbolic dynamics" in the general homoclinic situation. Here Smale assumed furthermore that the diffeomorphism f reduces to a linear one in some neighborhood of the point p at the expense of a suitable choice of a system of local coordinates–an assumption which, even though inessential from the point of view of "general position" (since one may arrange it by an arbitrarily small change in f), is nevertheless superfluous.

We return to the space $\Omega(N)$. Every sequence belonging to it can be recovered uniquely to within a shift if the lengths of the series of zeros separated by the blocks $\sigma = 1\ 2 \ldots (N-1)$ are given. Thus the elements of $\Omega(N)$ can be coded by sequences of natural numbers; namely,

$$\ldots n_{-1} n_0 n_1 n_2 \ldots \tag{16}$$

corresponds to the element

$$w = \cdots \sigma' \underbrace{00 \ldots 0}_{n_1} \sigma' \underbrace{00 \ldots 0}_{n_0} \sigma' \underbrace{0 \ldots 0}_{n_1} \sigma' \underbrace{0 \ldots 0}_{n_2} \sigma' \ldots .$$

If in $w = [\omega_n]$ all symbols $\omega_n = 0$ for $n > n^*$, then the sequence (16) will be bounded from the right (one may complete it with the symbol ∞), and in the same way it may be bounded from the left, or even consist of a finite number of symbols (if w consists only of a finite number of blocks σ'). We remark that if $\omega_n = 0$ for $n > n^*$, then $T^n w$ tends to a sequence consisting of zeros alone as $n \longrightarrow \infty$. The sequence consisting of zeros alone is a fixed point under the shift mapping and goes under the mapping φ into a fixed point p of the diffeomorphism f. It follows from this that to sequences (16) bounded from the right there correspond points whose trajectories are ω-asymptotic to p. Similarly, to sequences (16) bounded on the left (on both sides) there correspond points whose trajectories are α-asymptotic (doubly asymptotic) to p. A similar description (in terms of the sequences (16)) of trajectories remaining in a neighborhood of the closure of the trajectory of a homoclinic point has been put forward by L. P. Šil′nikov [61] (to be sure not for a diffeomorphism, but for a flow; but this is not very essential).(5) The above description in terms of TMC appears to me more responsive to the essence of the phenomenon, the more so as it can without difficulty be carried over to more involved situations.

PART III

QUASI-RANDOM OSCILLATIONS OF CERTAIN NONLINEAR SYSTEMS

§1. The pulsating one-dimensional potential well

We consider the simplest nonintegrable dynamical system, namely a one-dimensional oscillator in a periodically varying potential field. This system is described by a differential equation

$$\ddot{x} = -Q(x, t), \tag{1}$$

where $Q(x, t) = \partial U(x, t)/\partial x$ depends periodically on t. The choice of hypotheses regarding the dependence of Q on x has been determined by several considerations. On the one hand, these hypotheses must be sufficiently general that the essential nature of the process under consideration is clearly apparent. On the other hand, it was desired that the class under consideration should include an equation describing the motion in a particular case of the three-body problem. Finally, in the third place, certain hypotheses, in particular the choice of profile of the potential in the form of a symmetrical "well", simply facilitate calculations which would otherwise remain fairly formidable. These calculations are given in detail in [59] and [63] (see also [62]).

Our ultimate aim is the construction of a family of solutions of the equation (1), the investigation of which may be carried out by the methods of symbolic dynamics. Since, however, in Part II we were only concerned with mappings, our first step must be the construction of a cross-section in the sense of Poincaré and of the corresponding mapping (sequence function) S. Here one uses only the most general properties of the function Q, formulated below as "basic hypotheses". If supplementary hypotheses are fulfilled, one can construct for the mapping S an invariant set of fairly complicated structure, on which the action of S is isomorphic to a "topological Markov chain". Going back from the mapping S to the phase flow, we obtain the required "quasirandom" family of trajectories. The standard devices of symbolic dynamics now permit the construction of solutions of (1) with extremely varied behavior.

---

(5) And in fact for any open set V, containing the closure Γ of a transversal homoclinic curve, there exist a neighborhood $U \supset \Gamma$ and a number N^* such that the set of trajectories lying entirely in U is in one-to-one correspondence with the set of sequences (16) for which all $n_i \geqslant N^*$.

In Ju. I. Neĭmark's article [69] (see also [70]) a weaker assertion is given: for any neighborhood $U \supset \Gamma$ there exists an N^* such that to every sequence (16) with $n_i \geqslant N^*$ there corresponds a trajectory lying in U.

It is easy to show also that to each trajectory lying in some neighborhood $U_0 \supset \Gamma$ there corresponds some sequence (16). However, the possibility of choosing U_0 so that one obtains *all* sequences with the bounds $n_i \geqslant N^*$ and, above all, *only these*, does not follow from Neĭmark's arguments. In addition, in my opinion there are gaps in his proof.

And so we consider the motion of a material point (particle) in a field with

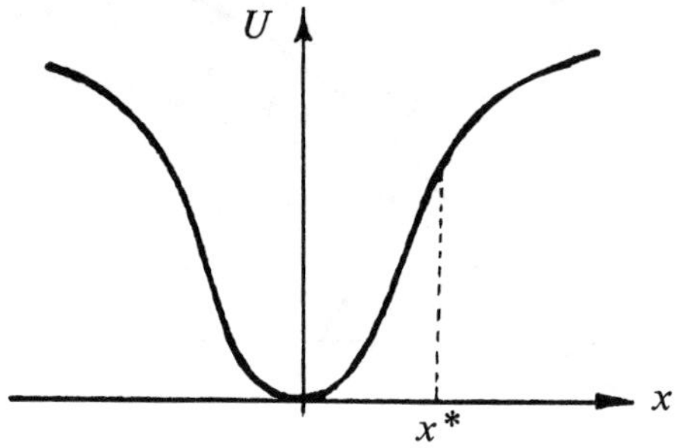

FIGURE 20

periodically varying potential $U(x, t)$, whose graph has the form of a symmetrical potential well (Figure 20). We assume that the force acting on the particle is always directed to the point $x = 0$ and far from this point falls off with the distance. The latter means that for large x the graph of U as a function of x is concave downwards. The profile of the potential well, as already stated, pulsates periodically, but its qualitative character remains one and the same. It is assumed also that the pulsation decays as $x \longrightarrow \infty$, and significantly more rapidly than the decrease in the force. Precise statements of the corresponding assumptions will be expressed for convenience in terms of the function $Q(x, t)$ (the force), and not $U(x, t)$ (the potential).

BASIC HYPOTHESES. A. $Q(x, t)$ is odd in x and 2π-periodic in t.

B. $Q > 0$ for $x > 0$.

C. $Q'_x \leqslant 0$ for $x > x^*$.

In view of A and B, the graph of a solution $x(t)$ of (1) as a function of t is concave downwards for $x > 0$, and upwards for $x < 0$ (Figure 21). Hence each solution of (1) has at least one zero, and consequently every trajectory of the phase-flow of the problem under consideration intersects the surface $x = 0$ at least once. This we take as the cross-section.

To construct the sequence function it is necessary that the trajectory should intersect the cross-section at least twice (Figure 22). Without supplementary hypotheses on Q, not all trajectories will in general have this property, and so the sequence function will be defined not on the whole cross-section and will be only a local diffeomorphism. We denote its domain by R^+, and the set of its values by R^-; R^+ consists of those points of the cross-section through which there pass trajectories which return to the surface under forward motion (in the sense of increasing time); through points of R^- pass trajectories which return to the surface under backward motion.

Let us consider what all this means in terms of a solution of (1). At the instant τ when a solution passes through zero, the state of the system is determined by a pair (v, τ), where $v = \dot{x}(\tau)$ (Figure 21). Since the dependence on time is assumed periodic, τ may be considered as a cyclic coordinate and, by the symmetry with respect to a change of sign in x, one may confine attention to nonnegative values of v. Finally, for

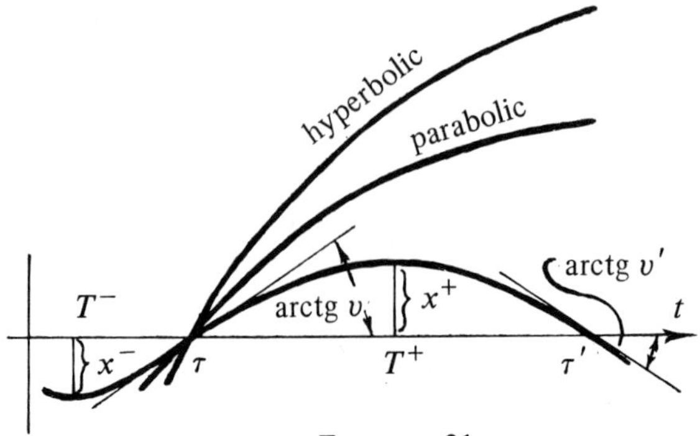

FIGURE 21

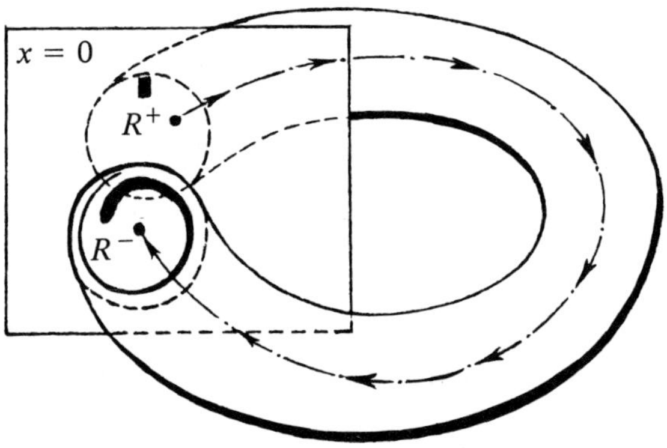

FIGURE 22

$v = 0$ the value of τ plays no part, since by the uniqueness theorem we can have $\dot{x}(\tau) = 0$ only for the trivial solution $x(t) \equiv 0$ (we recall that, by virtue of the oddness, $Q(0, t) \equiv 0$). This permits us to consider (v, τ) as polar coordinates in some plane Φ. (It would be more accurate to describe Φ as a quotient space obtained from the cross-section $x = 0$ with the aid of certain identifications, taking into account the indicated symmetries of the equation; we shall not dwell on this in greater detail.)

With (v, t) given, the solution $x(t)$ is determined uniquely (Figure 21). Let us see what can happen with a nontrivial solution for $t > \tau$. By the concavity in the region $x > 0$, the velocity $\dot{x}(t)$ is monotone decreasing. Hence two cases are possible. If $\dot{x}(t) > 0$ for all $t > \tau$, then the limit

$$v_\infty^+ = \lim_{t \to +\infty} \dot{x}(t) \tag{2}$$

exists. One may show, using only A and B, that in this case $x(t) \longrightarrow \infty$ as $t \longrightarrow +\infty$, even if $v_\infty^+ = 0$. If $v_\infty^+ > 0$, the solution is called *hyperbolic*; if $v_\infty^+ \equiv 0$, *parabolic*. The analogy with celestial mechanics (see Part I) is obvious.

If, however, there is an instant at which $\dot{x}$ vanishes (we denote by $T^+(v, \tau)$ the nearest such instant to τ on $(\tau, +\infty)$), then the increase of the solution $x(t)$ changes to decrease, and it returns to the value $x = 0$ at some instant τ' (again we use this notation for the instant on $(\tau, +\infty)$ nearest to τ when $x(\tau') = 0$) (Figure 21). In the phase space the corresponding trajectory returns to the cross-section (Figure 22) and, in consequence, one may define the sequence function S by the equality $S(v, \tau) = (v', \tau')$, where $v' = -\dot{x}(\tau')$ (the minus sign has appeared by way of convention; one considers only nonnegative values of v).

For $t > \tau'$, all these considerations are repeated; the solution may again turn out to be hyperbolic or parabolic or again return to zero at some instant τ''. In the last case on the pair (v, τ) a map $S^2: (v, \tau) \longrightarrow (v'', \tau'')$ is defined, where $v'' = \dot{x}(\tau'')$, and so on. For suitable initial conditions (v, τ) the solution $x(t)$ may have an infinite number of zeros, and so all powers of the mapping S will be defined. One sees by standard arguments in the theory of differential equations, based on the uniqueness theorem, that the zeros of a solution cannot have a point of accumulation.

Thus, subject to conditions A and B, the solutions of (1) may be classified as follows in regard to their behavior as $t \longrightarrow \infty$.

1) 0. The trivial solution $x(t) \equiv 0$.

2) Oscillating solutions having infinitely many zeros for $t > \tau$. These are in turn subdivided into:

B^+. Bounded oscillating.

OS^+. Unbounded oscillating (oscillatory).

3) Π. Parabolic solutions.

4) H. Hyperbolic solutions.

If one fixes the instant τ at which a solution vanishes, then the hyperbolic and parabolic solutions can be further subdivided according to the number of their zeros for $t > \tau$. The corresponding classes will later be denoted by Π_n^+ and H_n^+ ($n \geqslant 0$ being the number of zeros).

An analogous classification can be carried out for $t \longrightarrow -\infty$ also. The corresponding classes will be denoted by the same letters, with the index minus: B^-, H_n^-, and so on.

In the stationary case, when the force Q and the potential $U(x) = \int_0^x Q\,dx$ do not depend on the time, equation (1) is integrable by quadratures. Here the law of conservation of energy holds:

$$\tfrac{1}{2}\dot{x}^2 + U(x) \equiv h = \tfrac{1}{2}v^2 = \text{const.} \tag{3}$$

and the behavior of solutions is determined by the magnitude of the constant h. If $h < U(\infty) = \int_0^\infty Q(x)\,dx$, the solution will be periodic, and so will belong to B^+ or B^-. If however $h \geqslant U(\infty)$, then the solution will be monotone, will have exactly one zero and will tend to infinity as $t \longrightarrow \pm\infty$. For $h > U(\infty)$ the solution will be hyperbolic and will belong to $H_0^+ \cap H_0^-$, and for $h = U(\infty)$ it will be parabolic and belong to $\Pi_0^+ \cap \Pi_0^-$. Hence one has to consider separately two essentially different cases, according to whether $U(\infty)$ is finite or infinite. We call special attention to the fact that the behavior of the solutions as $t \longrightarrow -\infty$ and $t \longrightarrow +\infty$ is one and the same.

It is natural to suppose that the properties of solutions of the nonstationary equation (1) will be in some degree analogous to the properties of solutions of the averaged equation

$$\ddot{x} = -Q_0(x) = -\frac{1}{2\pi}\int_0^{2\pi} Q(x, t)\,dt. \tag{4}$$

In accordance with the above we distinguish two cases:

a) A potential well of infinite depth

$$J = \int_0^{\infty} Q_0(x)\,dx = \frac{1}{2\pi}\int_0^{2\pi}\int_0^{\infty} Q(x, t)\,dx\,dt = \infty.$$

In this case all nontrivial solutions of the averaged equation (4) are periodic.

b) A potential well of finite depth. In this case the equation (4) also has hyperbolic and parabolic solutions.

If we digress from the ideas of symbolic dynamics, our results concerning the equation (1) relate on the one hand to the behavior of its solutions as $t \longrightarrow \pm\infty$, and on the other to the question of the combinations in which the types of behavior as $t \longrightarrow -\infty$ and $t \longrightarrow +\infty$ can be linked up, i.e. in analogy to problems of evolution in celestial mechanics. These results are obtained under the conditions that the function Q, in addition to the basic assumptions A–C, satisfies a series of supplementary assumptions. In brief, one-sided behavior of solutions of (1) is analogous to the behavior of solutions of the averaged equation (4), but not the two-sided behavior: while the type of solutions of (4) as $t \longrightarrow -\infty$ and as $t \longrightarrow +\infty$ is one and the same, for (1) these types may be linked up in any manner.

TECHNICAL HYPOTHESES. D. Q, Q'_x, Q'_t and Q''_{tt} (the last only in the case $J = \infty$) are continuous for all x and t.

E. There exists a function $\psi(x)$, integrable over $[0, \infty)$, such that for all $x > 0$ and all t

$$|Q'_t(x, t)| \leqslant \psi(x), \qquad \text{if } J < \infty;$$

$$|xQ''_{tt}(x, t)| \leqslant \psi(x), \quad \text{if } J = \infty.$$

F. The ratio $\psi(x)/Q^2(x, t)$ is uniformly bounded for $x > x^*$ if $J < \infty$, and tends uniformly to 0 as $x \longrightarrow +\infty$ if $J = \infty$.

For the stationary equation, and so also the the averaged equation (4), there holds the law (3) of conservation of energy. In the nonstationary case energy is not conserved, but one can all the same construct an "energy-like" quantity $h^+(v, \tau)$, which enables one to classify solutions just as this is done with the aid of constant energy in the stationary case. The energy of a particle is comprised of kinetic and potential; this circumstance will serve us as a guiding concept. In fact, for a solution which returns to zero (i.e. such that $(v, \tau) \in R^+$), the velocity vanishes at the instant $t = T^+(v, \tau)$ (Figure 21), and the energy at this moment reduces to the potential term $U(X^+) = \int_0^{x^+} Q\,dx$, where $X^+(v, \tau)$ is the maximal value attained by the solution at the turning point. If however the solution is parabolic or hyperbolic (i.e. $(v, \tau) \notin R^+$), then the potential term tends to a constant $J = U(\infty) = \int_0^\infty Q\,dx$, and the energy is equal to $\frac{1}{2}(v_\infty^\pm)^2 + J$. In view of this we define the function

$$h^+(v,\tau) = \begin{cases} U_{\text{av}}(X^+) = \int_0^{X^+(v,\tau)} Q_0(y)\,dy = (2\pi)^{-1}\int_0^{2\pi}\int_0^{X^+(v,\tau)} Q(y,s)\,dy\,ds, & \text{if } (v,\tau) \in R^+; \\ \frac{1}{2}(v_\infty^\pm)^2 + J = \frac{1}{2}(v_\infty^\pm)^2 + \int_0^\infty Q_0(y)\,dy, & \text{if } (v,\tau) \in \mathrm{H}_0^+ \cup \Pi_0^+; \\ 0 & \text{if } v = 0. \end{cases}$$

LEMMA. *The function $h^+(v,\tau)$ is finite and continuous on the whole plane Φ, continuously differentiable on $\Phi\backslash\{0\}$ and has a unique minimum at the point* 0; *and its level lines $h^+(v,\tau) = C > 0$ are diffeomorphic to circles.*

Analogously, one defines a function $h^-(v,\tau)$ connected with the behavior of a solution as $t \longrightarrow -\infty$. In the stationary case these are obviously equal, so that $h^+(v,\tau) = h^-(v,\tau) = h = v^2/2$; but for an arbitrary equation (1) the functions h^+ and h^- need not coincide, which constitutes an extremely important fact.

COROLLARY. *If $J = \infty$, then all nontrivial solutions of* (1) *are oscillatory. If $J < \infty$, then the sets $\mathrm{H}_0^\pm$ and $R^\pm \cup \{0\}$ are open, and are respectively exterior and interior regions, bounded by smooth Jordan curves $\Pi_0^\pm$.*

All this follows at once from the lemma and the definition of the functions $h^\pm$. In particular,

$$\mathrm{H}_0^\pm = \{(v,\tau);\, h^\pm(v,\tau) > J\},$$

and

$$\Pi_0^\pm = \{(v,\tau);\, h^\pm(v,\tau) = J\}.$$

In the next section we prove that the regions $R^\pm \cup \{0\}$ have the same area. And since both of them contain the point 0, their bounding curves cannot lie one inside the other, and so intersect (see Figure 23).

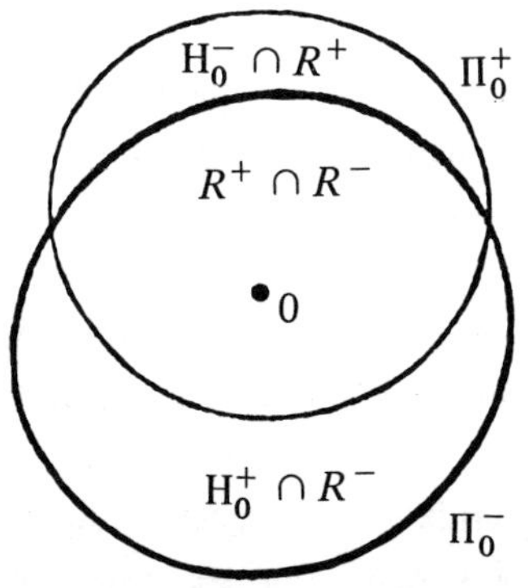

FIGURE 23

DEFINITION. If assumptions A–F hold, the equation (1) is called *typical*:

in the case $J < \infty$, if in at least one common point the curves Π_0^+ and Π_0^- have distinct tangents;

in the case $J = \infty$, if the equation $\psi(\tau) = \int_0^\infty xQ_t'(x, \tau)dx = 0$ has at least one simple root.

(The existence or roots for this equation is obvious, since ψ is a periodic function with zero mean value.)

THEOREM 1. *If assumptions* A–F *hold and the equation* (1) *is typical, then the following assertions are true*:

1) *There exists an N such that for every $n > N$ the equation* (1) *has a solution of period $4\pi n$ (of antiperiod $2\pi n$).*

2) *If $Q_0(x) \int_0^x Q_0(y)dy \longrightarrow 0$ as $x \longrightarrow +\infty$ (in particular, always if $J < \infty$), then all logically possible combinations of types of behavior of nontrivial solutions as $t \longrightarrow \pm\infty$ are realized, and for every combination, except $\Pi^+ \cap \Pi^-$ for $J < \infty$, the corresponding set of initial data has the cardinality of the continuum (in particular, there exist oscillatory solutions that are not bounded for $t > 0$ or for $t < 0$ or in both cases). All the sets $\Pi_n^+ \cap \Pi_m^+ \neq \varnothing$, since $\Pi^+ \cap \Pi^-$ is at least countable (this applies to $J < \infty$).*

3) *In the case $J < \infty$ solutions which tend to infinity on both sides can have any number of zeros. The set of initial data corresponding to solutions which are hyperbolic on both sides is open. The set of initial data corresponding to solutions which are oscillatory on one side and tend to infinity on the other, though not empty, has Lebesgue measure zero.*

We remark that if Q is sufficiently smooth and Q_t' is small, then by a theorem of Kolmogorov, Arnol'd and Moser the set of bounded oscillatory solutions is not nonempty, but has interior points.

As already mentioned, the hypotheses A–F were chosen so that the theory being developed should be applicable to a special case of the three-body problem, which will be considered in §3. However, one can pose a whole series of questions, so far unsolved, and undoubtedly presenting interest from the point of view of the general theory of nonlinear oscillations.

1) Is it possible to weaken the hypotheses E and F in such a way that the above corollary remains valid? The same question may be formulated separately for the cases $J = \infty$ and $J < \infty$. What role does assupmption C play in all this ?

2) Under what minimal conditions on Q does (1) still have a countable number of subharmonics? Can one here discard or weaken assumption C? Assumptions D–F?

3) In the statement of the theorem, can one discard or weaken the condition $Q_0(x) \int_0^x Q_0(y)dy \longrightarrow 0$?

4) Are the regions $R^\pm$ starlike with respect to the point 0? Is it true that $\partial h^\pm/\partial v > 0$ for $v > 0$? It is clear that a positive answer to the second question would imply a positive answer to the first. A negative answer to the first question would be

physically paradoxical, since then an increase in the initial velocity would lead to a hyperbolic motion changing to one returning to zero.

5) If Q has greater smoothness than envisaged in E, will the curves $\Pi_0^\pm$ be smoother? The same in the case of Q being analytic?

6) Do the typical equations form, in a suitable topology, an open everywhere dense set among all equations (1)?

7) What can be said about the measure of the set of unbounded oscillatory solutions? The same for the measure of the set of oscillatory bounded solutions for arbitrary Q (and not only nearly stationary Q)?

§2. Properties of the mapping S

In the previous section we constructed a mapping $S\colon R^+ \longrightarrow R^-$ whose domain R^+ coincides with $\Phi\backslash\{0\}$ for $J = \infty$, and lies inside the smooth curve Π_0^+ for $J < \infty$. On the region R^+ there are also defined two functions $T^+(v, \tau)$ and $X^+(v, \tau)$; we recall that $T^+(v, \tau)$ is the closest instant following τ when $\dot{x} = 0$, and $X^+(v, \tau)$ the value of $x(t)$ at this instant (Figure 21). Analogously, the following functions are defined on R^-: $T^-(v, \tau)$ is the nearest instant prior to τ when $\dot{x} = 0$, and $X^-(v, \tau) = |x(T^-(v, \tau))|$. The mapping $S\colon (v, \tau) \longrightarrow (v', \tau')$ corresponds to the transition from one zero of the solution $x(t)$ to the next (Figure 21); $T^+(v, \tau)$ lies between τ and τ', and plays the role of the instant T^-, for the pair (v', τ') of initial data, i.e.

$$T^+(v, \tau) = T^-(v', \tau'). \tag{5}$$

Hence

$$X^+(v, \tau) = X^-(v', \tau'), \tag{6}$$

and remembering that on $R^\pm$ we have by definition

$$h^\pm(v, \tau) = \int_0^{X^\pm(v,\tau)} Q_0(y)\,dy,$$

we conclude that $h^+(v, \tau) = h^-(v', \tau')$, or $h^+ = h^- \circ S$.

We consider the mappings $P^\pm\colon R^\pm \longrightarrow \Phi\backslash\{0\}$ defined in polar coordinates by the equations

$$P^\pm(v, \tau) = (X^\pm(v, \tau), T^\pm(v, \tau)).$$

It follows at once from standard theorems in the theory of differential equations that $P^\pm$ are diffeomorphisms, and that the diagram

$$\begin{array}{ccc} R^+ & \xrightarrow{\ S\ } & R^- \\ & {\scriptstyle P^+}\searrow \quad \swarrow{\scriptstyle P^-} & \\ & \Phi\backslash\{0\} & \end{array} \tag{7}$$

is, according to (5) and (6), commutative. The mappings S, P^+ and P^- may be continued to the point 0, which is a fixed point, though then they will be only homeomorphisms, since in general smoothness is lost at the point 0.

The equation (1) is equivalent to the canonical system

$$\dot{x} = p = \partial H/\partial p, \qquad \dot{p} = -Q(x, t) = -\partial H/\partial x, \tag{8}$$

with the Hamiltonian

$$H(p, x, t) = \tfrac{1}{2}p^2 + \int_0^x Q(y, t)\,dy. \tag{9}$$

The Poincaré-Cartan integral-invariant $\omega = p\,dx - H\,dt$ on the plane $x = 0$ assumes the form $\omega = -\tfrac{1}{2}p^2 dt$, and on Φ, where $p = v$ and $t = \tau$ are polar coordinates, $\oint_\gamma \omega$ is equal, with reversed sign, to the area bounded by the contour γ. Since $\oint \omega$ is preserved under a shift along a trajectory of the phase flow, and the mapping S is obtained in just this way, we have that S preserves area on Φ.

In the stationary case, when Q does not depend on the time, and, in particular, for the averaged equation (4), the mapping S looks fairly simple. From the energy integral (3) we find that

$$v' = v, \quad \tau' = \tau + 2\int_0^{v^2/2} \frac{du}{Q(x(u))\sqrt{v^2 - 2u}}, \tag{10}$$

where $x(u)$ is the function inverse to $U(x) = \int_0^x Q(y)\,dy$. Thus, the circles $v =$ const. are invariant under this mapping, and each of them rotates through its respective angle. If now we assume that the function $Q(x, t)$ is sufficiently smooth and almost stationary, then we find ourselves in the sphere of applicability of the theory of Kolmogorov, Arnol'd and Moser. According to this theory, the mapping S will have, also in the nonstationary case, invariant curves, close to circles. Let Γ be one of these curves. If the initial data (v, τ) are taken inside Γ, then the whole trajectory $\{S^n(v, \tau)\}$ will also lie inside Γ, and the corresponding solution $x(t)$ will be oscillatory and bounded. Hence the entire region interior to Γ belongs to $B^+ \cap B^-$, which clarifies the remark made in the last section after the statement of Theorem 1. Taking into account that S is area-preserving, we conclude that almost all points lying inside Γ, just as the solutions of equation (1) which they generate, are stable in the sense of Poisson (Poincaré's recurrence theorem).

If $J < \infty$, we then also deduce from the preservation of area that almost all solutions of (1), hyperbolic or parabolic as $t \to -\infty$ ($t \to +\infty$), will remain so for $t \to +\infty$ ($t \to -\infty$). This assertion is a consequence of the well-known theorem of Hopf ([65] or [66], Chapter 6), but it is nevertheless instructive to carry out its proof, since it is simple and graphic.

We shall denote the set of initial data (v, τ) in the plane Φ, which generate solutions of a given final type, by the same symbols as the types themselves, so that, for example, H_n^- or Π_n^- are the sets $(v, \tau) \in \Phi$, generating respectively hyperbolic or parabolic solutions of equation (1), having precisely n zeros for $t < \tau$. The set

$$A_n = (H_n^- \cup \Pi_n^-) \cap (B^+ \cup OS^+)$$

lies in R^+ and is transferred by the mapping S into A_{n+1} (why?). It follows from the definition that all the A_n are pairwise nonintersecting, and since they all have the same Lebesgue measure (a property of the mapping S) and lie in the bounded region R^+, it follows that the measure of each, and of all together, is equal to zero. It remains to remark that

$$(\mathrm{H}^- \cup \Pi^-) \cap (\mathrm{B}^+ \cup \mathrm{OS}^+) = \bigcup_{n=0}^{\infty} A_n.$$

Similarly, the measure of $(\mathrm{H}^+ \cup \Pi^+) \cap (\mathrm{B}^- \cup \mathrm{OS}^-)$ is zero also, as asserted in Theorem 1.

If the point (v_0, τ_0) is mapped by S into itself, i.e. if $v_0' = v_0$ and $\tau_0' = \tau_0 + 2\pi n$, then by the oddness of Q in x the solution $x(t)$ corresponding to this point will satisfy the identity

$$x(t + \tau_0) \equiv - x(t + \tau_0').$$

We see from this that $2\pi n = \tau_0' - \tau_0$ is for $x(t)$ an antiperiod $(x(t + 2\pi n) = -x(t))$, and $4\pi n$ is a period. A simple calculation shows that

$$T^+(v_0, \tau_0) - T^-(v_0, \tau_0) = 2\pi n. \tag{11}$$

Using an idea which goes back to Poincaré, one may attempt to prove the existence of such periodic solutions with the aid of the following considerations. We consider a region G consisting of the point 0 and those points of $R^- \cap R^+$ at which

$$T^+(v, \tau) - T^-(v, \tau) < 2\pi n. \tag{12}$$

For sufficiently large n both G and SG are nonempty, contain the point 0, and have the same area, and so their boundaries Γ and $S\Gamma$ have common points. Hence they have a common point (X, T) and curves $\gamma = P^-\Gamma$ and $\gamma' = P^-S\Gamma = P^+\Gamma$.

We put $(v_0, \tau_0) = (P^-)^{-1}(X, T)$. Then $(X, T) = (X^-(v_0, \tau_0) T^-(v_0, \tau_0))$, and since $(v_0, \tau_0) \in (P^-)^{-1}\gamma = \Gamma = \partial G$ and (11) must hold on G, it follows that $T^+(v_0, \tau_0) = T^-(v_0, \tau_0) + 2\pi n = T + 2\pi n$. Hence, the point

$$P^+(P^-)^{-1}(X, T) = P^+(v_0, \tau_0) = (X^+(v_0, \tau_0), T^+(v_0, \tau_0))$$

lies on the same ray $T =$ const. as the point (X, T).

We assume now that *the curve γ' is starlike with respect to the point* 0, i.e. that every ray $T =$ const. intersects γ' in precisely one point. Then both points $(X, T) \in \gamma \cap \gamma'$ and $P^+(P^-)^{-1}(X, T) \in P^+\Gamma = \gamma'$ must coincide: $P^+(P^-)^{-1}(X, T) = (X, T)$, whence, by (7),

$$S(v_0, \tau_0) = S(P^-)^{-1}(X, T) = (P^-)^{-1}P^+(P^-)^{-1}(X, T) = (P^-)^{-1}(X, T).$$

Hence the desired fixed point has been found.

Unfortunately, I have not succeeded in proving the above assumption of the starlike character of γ' under the general assumptions A–F concerning the function Q. For each fixed n one may ensure it by requiring that Q is almost stationary, since in the stationary case the curves (11) and their images under the mapping P^+ are circles. It is not clear whether one can do the same for infinitely many n at the same time. Hence for the proof of the existence of an infinite number of subharmonics of period $4\pi n$, as this was enunciated in Theorem 1, the above argument had to be somewhat altered.

We take in place of the curve Γ defined by (11) (it is unclear, by the way, whether Γ can consist of several components; this question also, like that of the star-likeness of γ', is related to Problem 4, posed at the end of §1), the family of curves $\Gamma_c = \{(v, \tau); h^+(v, \tau) = c\}$. By the lemma of §1, these curves are diffeomorphic to circles, and

$$S\Gamma_c = \{(v, \tau), h^-(v, \tau) = c\}.$$

Recalling the definition of the function h^-, we see that $X^-(v, \tau) = c_1$ on the curve $S\Gamma_c$, where c_1 is determined by the relation $\int_0^{c_1} Q_0(y)\,dy = c$, and so $\gamma'_c = P^- S\Gamma_c = P^+\Gamma_c$ is a circle with center 0, i.e. the condition of starlikeness is fulfilled trivially.

If now one can find a c such that (11) holds for some point of $\Gamma_c \cap S\Gamma_c$ (this set is nonempty, since S preserves area), then all the above arguments remain in force and this point will be fixed under the mapping S.

The points of $\Gamma_x \cap S\Gamma_c$ are determined by the equation

$$h^+(v, \tau) = h^-(v, \tau) = c, \tag{13}$$

and, if the equation (1) is typical in the sense of §1, there exists an arc d: $[c_0, J) \longrightarrow R^- \cap R^+$ such that $(v(c), \tau(c)) = d(c) \in \Gamma_c \cap S\Gamma_c$. In the case $J < \infty$ this follows at once from the implicit function theorem, applied to a neighborhood of a transversal point of the intersection of the curves $\Pi_0^+ = \{(v, \tau)\colon h^+(v, \tau) = J\}$ and $\Pi_0^- = \{(v, \tau)\colon h^-(v, \tau) = J\}$. In the case $J = \infty$, one uses the asymptotics of $h^\pm(v, \tau)$ as $v \longrightarrow \infty$, in which an essential role is played by the function $\psi(\tau) = \int_0^\infty xQ'_t(x, \tau)\,dx$.

From assumptions A–F one may deduce that along the arc d we have $T^* - T \longrightarrow \infty$ as $c \longrightarrow J$. Hence for all large n there are points on it where (11) holds, which completes the proof.

We take up now, in somewhat greater detail, properties of the mapping S in the case of a potential well of finite depth ($J < \infty$). As already stated, $\Pi_0^+ \cap \Pi_0^- \neq \varnothing$, and in the simplest "typical" case these curves intersect in two points (Figure 23), dividing Φ into four regions. The behavior of solutions $x(t)$ with initial data (v, τ) taken from these regions is represented in Figure 24, a–d.

To points (v, τ) lying outside both curves there corresponds a solution having a unique zero $t = \tau$ and hyperbolic on both sides (Figure 24a). Points of the "tongue" $R^+ \cap H_0^-$ lying inside Π_0^+ but outside Π_0^- correspond to solutions hyperbolic for $t < \tau$, but having at least one more zero for $t > \tau$ (Figure 24b). The mapping S takes R^+ into R^-, and to the extent of approximation to $\Pi_0^+ = \partial R^+$ the time of return $\tau' - \tau$ (i.e. the distance between the zeros in Figure 24b) tends to infinity. At the limits of the "tongue" $R^+ \cap H_0^-$ one may consider τ bounded, if one fixes one of the branches of the polar angle on the plane Φ. Then $\tau' \longrightarrow \infty$ on approach to Π_0^+, and so the image

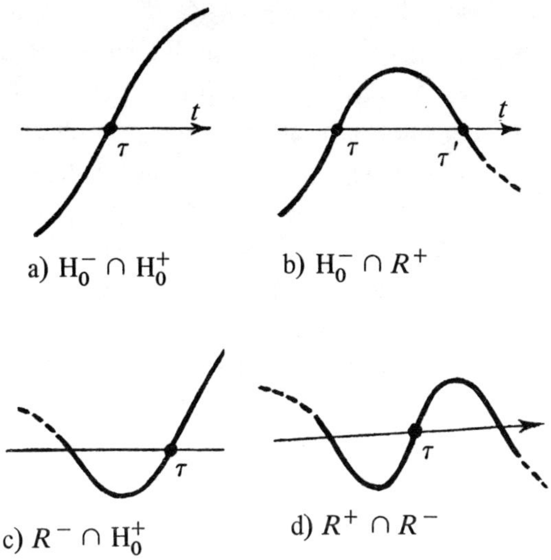

FIGURE 24

of the "tongue" under the mapping S will be a spiral-formed "snake", winding its "tail" forever on the curve Π_0^+ (Figure 25).

It is not hard to see that initial data (v, τ) lying in the "tail" generate solutions of type H_1^-, i.e. having for $t < \tau$ precisely one zero and hyperbolic as $t \longrightarrow -\infty$ (Figure 26a). The difference between this picture and Figure 24b) arises, on the one hand, from our agreement to consider only $v \geqslant 0$ (which led to a mirror image in $0X$),

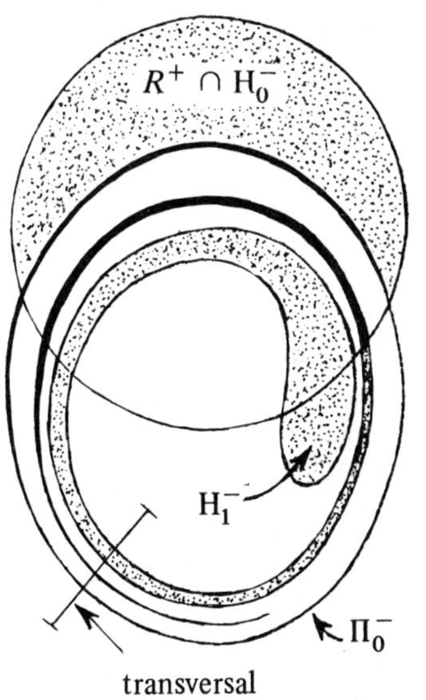

FIGURE 25

and, on the other, from the fact that we now have τ where previously τ' stood, corresponding to carrying out the mapping S: $H_0^- \cap R^+ \longrightarrow H_1^-$.

Part of the "tail" (Figure 25), which is outside Π_0^+, corresponds to solutions H_0^- and H_0^+ (Figure 26b), and that part which is inside Π_0^+ corresponds to solutions which have at least three zeros (Figure 26c). (For simplicity we do not mention, here and in the sequel, limiting cases of solutions, parabolic in one or both directions.) The intersection $H_1^- \cap R^+$ contains a countable number of components, which accumulate towards the arc of the curve Π_0^-, lying inside Π_0^+. The mapping S takes $H_1^- \cap R^+$ into H_2^-, which also consists of a countable number of components. Except, possibly, for a finite number, each of the components consists of a double spiral, between whose branches lies a "serpent" H_1^-. Both branches together with the "tail" wind onto Π_0^+ (Figure 27). The sequence of these double spirals accumulate to the boundary of the serpent H_1^-, i.e. to the curve Π_1^-.

This construction is repeated indefinitely. In the end, one obtains the following representation of the formation of the sets $H^- = \bigcup_0^\infty H_n^-$. Each of the sets H_n^-, $n \geqslant 2$, lies inside Π_0^- and consists of a countable number of components, of which a finite number may be formed irregularly, while the remainder form double spirals, covering the "serpent" H_1^-, as in Figure 27, and accumulating to Π_0^- from inside. Each such component is bounded by two curves, which are components of Π_n^-; these

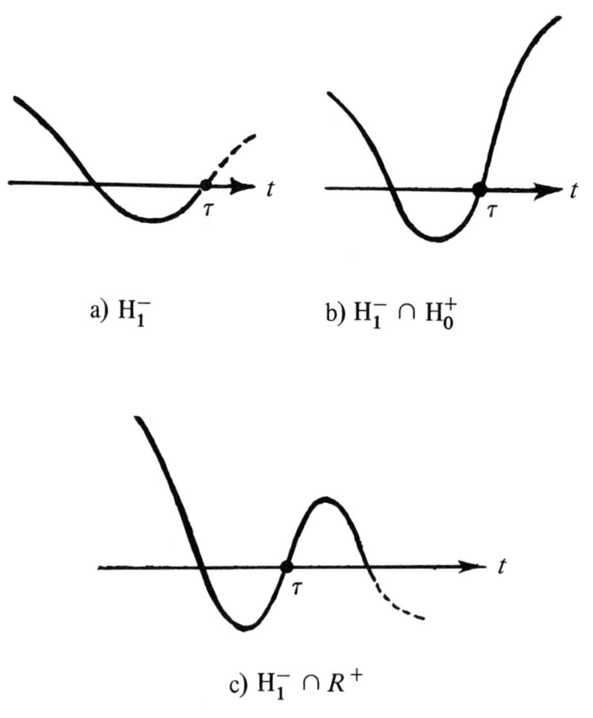

FIGURE 26

curves are diffeomorphic to a line, and accumulate at both ends to Π_0^-. Finally, each of these curves is the limit of a countable sequence of components of H_{n+1}^-.

It is clear that the complement

$$\Phi \setminus \bigcup_{n=0}^{\infty} (H_n^- \cup \Pi_n^-) = B^- \cup OS^- \cup \{0\}$$

has a fairly complicated structure. Its structure may be in part understood by considering the intersection with a transversal to the curve Π_0^- (Figure 25). The function $\xi = h^-(v, \tau) - J$ constitutes a coordinate on the transversal, at least in the neighborhood of the point of intersection with Π_0^-, with this point corresponding to $\xi = 0$. Points at which $\xi > 0$ belong to H_0^-, and the interval $(-\epsilon, 0)$ lies in R^- (Figure 28). For sufficiently small ϵ, the trace of the set $B^- \cup OS^- \cup \Pi^-$ on the interval $[-\epsilon, 0]$ of the

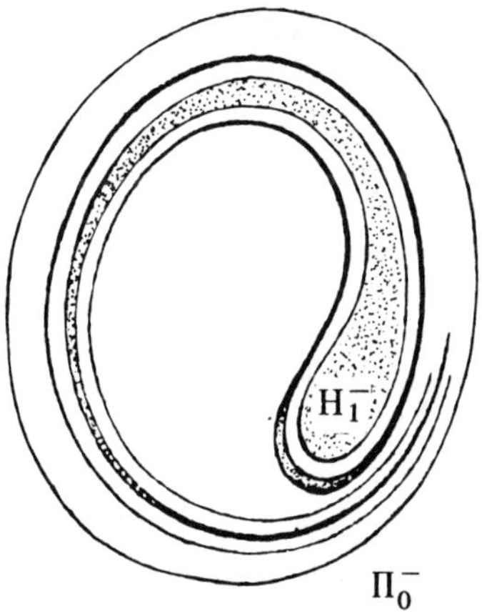

FIGURE 27

transversal is obtained with the aid of a construction analogous to the construction of the Cantor set. In the first stage one removes from the interval a countable number

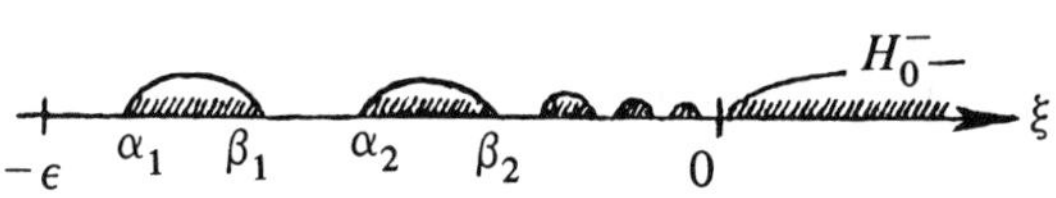

FIGURE 28

of intervals $(\alpha_{n_1}, \beta_{n_1})$ converging to 0. Then from each interval $[\beta_{n_1}, \alpha_{n_1+1}]$ one removes a countable number of intervals $(\alpha_{n_1 n_2}, \beta_{n_1 n_2})$ converging to β_{n_1} as $n_2 \to -\infty$, and to α_{n_1+1} as $n_2 \to +\infty$. Furthermore, from each remaining interval one removes a countable family of intervals $(\alpha_{n_1 n_2 n_3}, \beta_{n_1 n_2 n_3})$, and so on. Each of the discarded intervals belongs to one of the components of a set H_m^-, and its ends belong to Π_m^-. After discarding all the intervals, there remains a set of zero measure, which forms the trace of $OS^- \cup B^- \cup \Pi^-$ on an interval of the transversal. Accordingly, near Π_0^- the whole set $OS^- \cup B^- \cup \Pi^-$ has the local structure of the product of a zero-dimensional compact set (i.e. in the finite power of a Cantor set) with a one-dimensional interval. The whole picture near Π_0^- reminds one to some degree of a layer cake, in which the hyperbolic regions are distributed as layers of frosting, bordered with parabolic curves, between which are concealed a continuum of sheets of oscillating and bounded motions.

An entirely similar structure is possessed by the sets $\mathrm{H}^+ = \bigcup_0^\infty \mathrm{H}_n^+$, $\Pi^+ = \bigcup_0^\infty \Pi_n^+$ and $\mathrm{OS}^+ \cup \mathrm{B}^+$, except that in this case the spirals are wound around the curve Π_0^+.

To explain the possible link-up between types of behavior of solutions as $t \longrightarrow -\infty$ and as $t \longrightarrow +\infty$, it is necessary to place both spiral layers of the cake one on the other. In the typical case, when Π_0^- and Π_0^+ intersect transversely, this placement in some neighborhood U of the points of intersection takes place regularly, in the sense that each sheet of the one structure intersects with each sheet of the other structure, falling in the same neighborhood U. Hence one obtains without difficulty all the rest of the assertions of Theorem 1. One can, however, go further. In fact, in a neighborhood of points of a transversal intersection of Π_0^- and Π_0^+ one can distinguish an invariant set on which the action of S is described in the terms of "symbolic dynamics."

Let $P = (v_P, \tau_P)$ be one such point. In a sufficiently small neighborhood one may take as coordinates

$$\xi = h^-(v, \tau) - J, \qquad \eta = h^+(v, \tau) - J. \tag{14}$$

The values $\xi = \eta = 0$ correspond to the point P itself; the corresponding solution of (1) has a unique zero at $t = \tau_P$ and is parabolic as $t \longrightarrow \pm\infty$. We denote this solution by $x_P(t)$, and will need it again in the sequel. We take the neighborhood U to be given by the inequalities $|\xi|, |\eta| \leqslant \epsilon$. The lower half of U belongs to R^+ by the inequality $h^+(v, \tau) < J$ (see §1), and so the mapping S is defined in it. On the set $U \cap S^{-1}U$ this map may be written in coordinates (14). Taking into account that $h^+ = h^- \circ S$, we have

$$\xi' = \eta, \qquad \eta' = f(\xi, \eta). \tag{15}$$

We can find and estimate the derivative f'_ξ from the condition of the preservation of area, to wit:

$$\frac{v'}{v} = \frac{\partial(v', \tau')}{\partial(v, \tau)} = \frac{\partial(v', \tau')}{\partial(\xi', \eta')} \begin{vmatrix} 0 & 1 \\ f'_\xi & f'_\eta \end{vmatrix} = -f'_\xi \cdot \frac{Z(v, \tau)}{Z(v', \tau')},$$

where $Z(v, \tau) = \partial(\xi, \eta)/\partial(v, \tau) = \partial(h^-, h^+)/\partial(v, \tau)$. For a sufficiently small neighborhood U the ratios v'/v and $Z(v', \tau')/Z(v, \tau)$ will be close to 1 in $U \cap S^{-1}U$, and so $f'_\xi \approx -1$. In the sequel we will take it that $|f'_\xi| \leqslant 2$.

If $|f'_\eta| > 1 + |f'_\xi|$, then the differential of the mapping (15) will be a hyperbolic linear map, i.e. will have real eigenvalues, one of modulus less than 1, and one of modulus greater than 1. With the above convention as to the choice of the neighborhood U, this means that $|f'_\eta|$ must be greater than 3. However for this, in order to be able to carry out certain majorizing bounds, we need that $|f'_\eta|$ should be large "with a margin", in fact that in $U \cap S^{-1}U$ for some $a \in (0, \epsilon)$ the inequality

$$\frac{1 + \max|f'_\xi|}{\min|f'_\eta|} < \frac{\epsilon - a}{\epsilon + a} \tag{16}$$

should hold.

The intersection $U \cap S^{-1}U$ is certainly not empty. For on the arc of the curve $h^-(v, \tau) = h^+(v, \tau)$, which in the coordinates (14) is a part of the bisectrix $\xi = \eta$

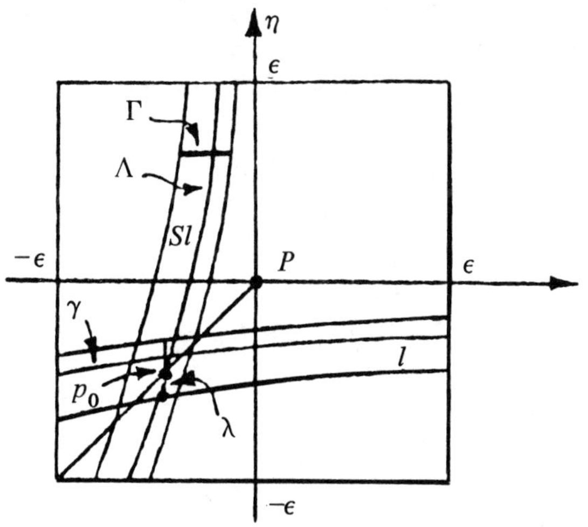

FIGURE 29

(Figure 29), there lies a sequence of fixed points of the mapping S, converging to $P = (0, 0)$. Starting with a certain one, all will lie in U, and so in $U \cap S^{-1}U$.

Let $p_0 = (\xi_0, \xi_0)$ be one of these fixed points for which $|\xi_0| < a$. We denote by l a component of $U \cap S^{-1}U$ containing this point. Our first aim is to show that l and Sl are distributed in U "regularly" (see Figure 18 in Part II).

We draw through p_0 a segment λ of a vertical line up to the intersection with the boundary of l; we denote the image $S\lambda$ by Λ (Figure 29). Since λ and Λ lie in U, the time of return $\tau' - \tau$ for the solution of (1) with initial data belonging to λ remains bounded (for a periodic solution, determined by the point p_0, this time is equal to $2\pi n$, and for others it cannot differ from $2\pi n$ by more than twice the oscillation of the polar angle in the neighborhood U. Hence the upper bounding point of the segment λ cannot lie on the line $\eta = 0$, i.e. on Π_0^+, since on approach to Π_0^+, as already discussed in § 1, the time of return tends to infinity. Since $\xi' = \eta$ (see (15)), it follows that Λ also is separate from the line $\xi = 0$ and lies strictly to the left of it. From the relation

$$\eta' - \xi_0 = f(\xi_0, \eta) - f(\xi_0, \xi_0) = f'_\eta(\xi_0, c)(\eta - \xi_0), \tag{17}$$

the hypothesis $|\xi_0| < \epsilon$ and (7) we find that on λ

$$\eta = \xi_0 + \frac{\eta' - \xi_0}{f'_\eta} > \xi_0 - \frac{\epsilon + a}{|f'_\eta|} > \xi_0 - (\epsilon - a) > -\epsilon, \tag{18}$$

and so the whole segment together with its ends lies in the interior of $U \cap R^+$. For Λ this means the impossibility of exit on the vertical sides of the rectangle $-\epsilon \leqslant \xi \leqslant 0$, $|\eta| \leqslant \epsilon$, and so the ends of Λ must lie on the lines $\eta = \pm\epsilon$. Hence Λ must intersect every horizontal line, and from (17) it follows that this must be in precisely one point, since otherwise, in disagreement with (16), f'_η would vanish. Through a point $(\xi, \eta_1) \in \Lambda$ we draw a segment Γ of a horizontal line up to the intersection with the boundary of $S(l)$, and we denote by γ its preimage, $\gamma \subset l$. By the same arguments as above, we establish that γ intersects each vertical line in U precisely once, and that its ends lie on the lines $\xi = \pm\epsilon$; it is only essential to verify that γ cannot intersect the line $\eta = -\epsilon$. Along γ, according to (15), $f(\xi, \eta) = \text{const}$. Hence, taking the point (ξ_0, η_1) as the intersection $\lambda \cap \gamma$, we find that

$$0 = f(\xi, \eta) - f(\xi_0, \eta_1) = \widetilde{f}'_\xi(\xi - \xi_0) + \widetilde{f}'_\eta(\eta - \eta_1),$$

$$\eta > \eta_1 - \frac{|\widetilde{f}'_\xi|}{|\widetilde{f}'_\eta|}|\xi - \xi_0| > \eta_1 - \frac{|\widetilde{f}'_\xi|}{|\widetilde{f}'_\eta|}(a + \epsilon)$$

$$> \xi_0 - \frac{\epsilon + a}{|\widetilde{f}'_\eta|} - \frac{|\widetilde{f}'_\xi|}{|\widetilde{f}'_\eta|}(\epsilon + a) > -a - \frac{1 + \max|f'_\xi|}{\min|f'_\eta|}(\epsilon + a) > -\epsilon$$

in accordance with (16) and (18).

One may now verify without difficulty that the component l is a curvilinear quadrilateral bounded by segments of the lines $\xi = \pm\epsilon$ and the two curves γ, constructed for $\eta_1 = \pm\epsilon$. Similarly, Sl is bounded by segments of lines $\eta = \pm\epsilon$, and two arcs, images of the vertical sides of l, which intersect each horizontal line $\eta = \eta_1$, $|\eta| \leqslant \epsilon$, in precisely one point.

As already mentioned, for solutions with initial data from l, the time of return $\tau' - \tau$ differs from $2\pi n$ (i.e. from the return time of a periodic solution corresponding to the fixed point p_0) by not more than twice the oscillation of the polar angle in U. Hence, if the neighborhood U is sufficiently small, then l cannot contain two fixed points with different times of return (for fixed points it is always a multiple of 2π). And since to each $n \geqslant N$ there corresponds a fixed point with the return time $2\pi n$ and for sufficiently large N they all satisfy the condition $|\xi_0| < a$, it follows that $U \cap S^{-1}U$ contains a countable number of components l_n of the type described above (in the sequel we shall term them regular); in Figure 30 two of them are indicated.

We can now begin to carry through arguments, similar to those used in connection with "Smale's horseshoe" and its generalizations, considered in Part II. To be sure, the mapping $S\colon l \longrightarrow Sl$ is now nonlinear, but it nevertheless enjoys the key property, namely a strong magnification in one direction (nearly vertical) and a strong contraction in another (nearly horizontal). A detailed justification of the subsequent arguments may be found in [59], Part I.

The image Sl_n of every regular component intersects every regular component l_m. Hence, as for the "horseshoe", one deduces that for any sequence $w = [m_n; -\infty < n < +\infty, m_n \geqslant N]$ the intersections

$$\varphi(w^+) = \bigcap_{n=0}^{\infty} S^{-n} l_{m_n}, \qquad \varphi(w^-) = \bigcap_{n=-\infty}^{-1} S^{-n} l_{m_n},$$

$$\varphi(w) = \bigcap_{n=-\infty}^{+\infty} S^{-n} l_{m_n} \tag{19}$$

are nonempty. The first of them is extended along l_{m_0} and is the graph of a continuously differentiable function $\eta = f(\xi)$; the second is extended along $Sl_{m_{-1}}$ and is the graph of a function $\xi = g(\eta)$ (also of class C^1); finally, $\varphi(w)$ is a point.

Denoting by Ω_N^∞ the space of all sequences w equipped with the natural product topology, we obtain a mapping $\varphi\colon \Omega_N^\infty \to R^+ \cap R^-$, which is a homeomorphism. As always, Ω_N^∞ admits the shift homeomorphism T to the left by one symbol, and the diagram

$$\begin{array}{ccc} \Omega_N^\infty & \xrightarrow{T} & \Omega_N^\infty \\ \varphi \downarrow & & \downarrow \varphi \\ R^+ & \xrightarrow{S} & R^- \end{array} \tag{20}$$

is commutative. In particular, all powers of the mapping S are defined on $\varphi(\Omega_N^\infty)$.

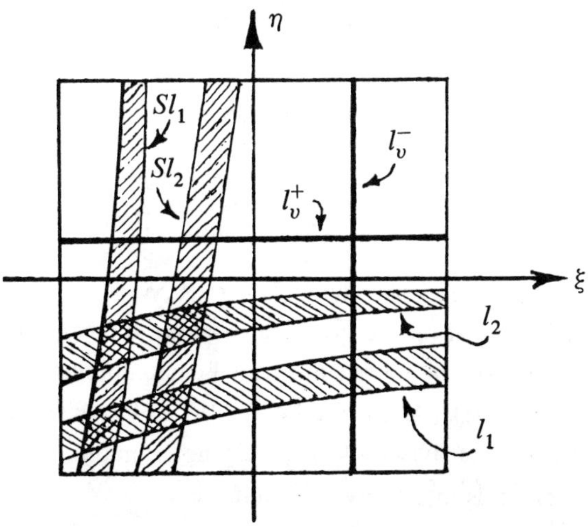

FIGURE 30

We recall now that to points $(v, \tau) = (\xi, \eta)$ lying in the region $\eta \geqslant 0$ there correspond solutions that do not vanish for $t > \tau$ and tend to infinity as $t \longrightarrow +\infty$. From the definition of $h^+(v, \tau)$ and (14) we obtain the velocity at infinity:

$$v_\infty^+ = \lim_{t \to +\infty} \dot{x}(t) = \sqrt{2h^+(v, \tau)} = \sqrt{2\eta}\ . \tag{21}$$

We put $\delta = \sqrt{2\epsilon}$ and for $0 \leqslant \nu < \delta$ denote by l_ν^+ the horizontal line segment given by $\eta = \nu^2/2$, $|\xi| \leqslant \epsilon$. According to (21), for all solutions with initial data from l_ν^+ the velocity at infinity is precisely equal to ν. Similarly, by l_ν^- we denote the vertical line segment $\xi = \nu^2/2$, $|\eta| \leqslant \epsilon$. Solutions with initial data $(v, \tau) \in l_\nu^-$ do not vanish for $t < \tau$, and for them $v_\infty^- = \lim_{t \to -\infty} |\dot{x}(t)| = \nu$.

The image Sl_m of every regular component intersects every segment l_ν^+, and every regular component l_m intersects every segment l_ν^- (Figure 30). Hence l_ν^+ is included in "horseshoe" arguments equally with l_m, and l_ν^- equally with Sl_m. For example, the intersection

$$d = l_{m_0} \cap S^{-1} l_{m_1} \cap \cdots \cap S^{-(n-1)} l_{m_{n-1}} \cap S^{-n} l_\nu^+ \tag{22}$$

is nonempty, lies on l_{m_0} and is the graph of a function $\eta = f(\xi)$. On d the mappings $S, S^2, \ldots, S^n$ are defined. But $S^n d \subset l_\nu^+$ lies outside R^+, and so S^{n+1} is not defined. Hence for $t > \tau$ a solution with initial data $(v, \tau) \in d$ has exactly n zeros (we recall that the mapping S is defined by a shift along the solution from one zero to the next) and then goes to infinity with velocity ν. (The reader is recommended to work out, using Figure 30, all stages of the construction of the intersection (22), for example with $n = 3$.)

We extend the space Ω_N^∞, taking into consideration all possible sequences $w = [\omega_n, n_1 \leqslant n \leqslant n_2]$, where $-\infty \leqslant n_1 < 0 \leqslant n_2 \leqslant +\infty$ and the symbols ω_n are natural numbers $\geqslant N$ for $n_1 < n < n_2$ and real numbers from the segment $[0, \delta]$ for $n = n_1$ or n_2. We denote the space thus obtained (though so far it is just a set) by $\Omega(N, \delta)$. In addition to Ω_N^∞, it contains sequences bounded from the left, from the right, or from both sides:

$$\begin{gathered} \nu^- m_{n_1+1} m_{n_1+2} \cdots m_{-1} m_0 \cdots m_n \cdots, \\ \cdots m_n \cdots m_{-1} m_0 \cdots m_{n_2-1} \nu^+, \\ \nu^- m_{n_1+1} \cdots m_{-1} m_0 \cdots m_{n_2-1} \nu^+. \end{gathered} \tag{23}$$

We equip the set of admissible symbols $L = [0, \delta] \cup \{m; m \geqslant N\}$ with a topology that coincides with the usual one on $[0, \delta]$ and makes the natural numbers m isolated points that converge to 0 as $m \longrightarrow \infty$ (thus, L is homeomorphic to the subset $[0, \delta] \cup \{-1/m; m \geqslant N\}$ of the line). A neighborhood $U(w^0)$ of an element $w^0 = [\omega_n^0; n_1^0 \leqslant n \leqslant n_2^0]$ is given by a finite subset $\nu \subseteq [n_1^0, n_2^0]$ not containing $\pm\infty$ and neighborhoods $V(\omega_n^0) \subseteq L$ of the symbols ω_n^0, $n \in \nu$. By definition,

$$U(w^0) = \{w = [\omega_n, n_1 \leqslant n \leqslant n_2]; \nu \subseteq [n_1, n_2], \omega_n \in V(\omega_n^0) \text{ for } n \in \nu\}.$$

It is easy to see that in this topology the convergence

$$w^{(k)} = [\omega_n^k; n_1^k \leqslant n \leqslant n_2^k] \longrightarrow w^0$$

holds if and only if

a) $\overline{\lim}\, n_1^k \leqslant n_1^0, \quad n_2^0 \leqslant \underline{\lim}\, n_2^k$, and

b) $\lim \omega_n^k = \omega_n^0$ for every $n \in [n_1^0, n_2^0]$ distinct from $\pm\infty$

(the latter means, in particular, that for $\omega_n^0 = m$ all ω_n^k, starting at some point, are equal to m, and that $\omega_n^k = m_k \longrightarrow \infty$ for $\omega_n^0 = 0$). The space $\Omega(N, \delta)$ is compact.

The convention $n_1 < 0 \leqslant n_2$, adopted in the definition of admissible sequences of symbols, is not insignificant; the homeomorphism T of the left shift is now defined not on all sequences $w = [\omega_n: n_1 \leqslant n \leqslant n_2]$, but only on those for which $n_2 > 0$ (and so ω_0 must be a natural number $\geqslant N$). We denote the domain of T by Δ^+. Accordingly, $\Delta^- = T\Delta^+$ consists of those w for which $n_1 < -1$.

The domain of definition R^+ of our mapping $S: R^+ \longrightarrow R^-$ likewise does not coincide with the whole plane Φ, and so the concepts "trajectory" and "invariant set" need to be made more precise. We shall assume the trajectory $\{S_p^n\}$ of a point p to be the maximal continuation on both sides (so long as the iterations S_p^n are defined at all); we call a set M *invariant* if $S(M \cap R^+) = M \cap R^-$. Let V be an arbitrary neighborhood. The set of all points p whose trajectories lie wholly in V will be a maximal invariant set contained in V; we will denote it by M_V.

We now extend the definitions (19) to all $w = [\omega_n; n_1 \leqslant n \leqslant n_2] \in \Omega(N, \delta)$, as follows:

$$\varphi(w^+) = \bigcap_{n=0}^{n_2} S^{-n} l_{\omega_n}, \quad \varphi(w^-) = \bigcap_{n=n_1}^{-1} S^{-n} l_{\omega_n},$$

$$\varphi(w) = \bigcap_{n=n_1}^{n_2} S^{-n} l_{\omega_n}. \tag{24}$$

For infinite n_1 and n_2 these formulas coincide with (19). If $n_2 < +\infty$, then w terminates (cf. (23)) with a number $V^+ \in [0, \delta]$, and $l_{\omega_{n_2}}$ means in (24) a vertical segment $l_{V^+}^+ \subset \{\eta = \frac{1}{2}(V^+)^2\}$ (Figure 30). If again $n_1 > -\infty$ and w starts with a number V^-, then one must then interpret (24) in a somewhat nonstandard manner: $S^{|n_1|} l_{\omega_{n_1}}$ is replaced by $S^{|n_1|-1} l_{V^-}^-$, where $l_{V^-}^- \subset \{\xi = \frac{1}{2}(V^-)^2\}$ is a vertical segment (we recall that l_V^- is distributed in the neighborhood U similarly to the sets Sl_m, which explains the irregularity in the formulas).

THEOREM 2 ([59], Part III). *If conditions* A–F *of* §1 *are fulfilled,* $J < \infty$, *and the equation* (1) *is typical, i.e.* Π_0^- *and* Π_0^+ *intersect transversally in some point P, then there exist N,* δ *and a neighborhood V of the point P satisfying the following conditions*:

1) *The mapping* $w \longrightarrow \varphi(w)$ *is a homeomorphism between* $\Omega(N, \delta)$ *and the maximal invariant set* M_V *in the neighborhood* V, $\varphi(\Delta^{\pm}) = R^{\pm} \cap M_V$, *and the diagram*

$$\begin{array}{ccc} \Delta^{+} & \xrightarrow{T} & \Delta^{-} \\ \Big\downarrow{\scriptstyle\varphi} & & \Big\downarrow{\scriptstyle\varphi} \\ M_V \cap R^{+} & \xrightarrow{S} & M_V \cap R^{-} \end{array}$$

is commutative.

2) *In the coordinates* (14) *the sets* $\varphi(w^{+})$ *and* $\varphi(w^{-})$ *are, respectively, the graphs of functions* $\eta = f(\xi)$ *and* $\xi = g(\eta)$ *of class* C^1; *the tangents to them determine on* M_V *the structure of a hyperbolic set*; $\varphi(w^{+})$ *is locally stable for the point* $p = \varphi(w)$, *and* $\varphi(w^{-})$ *is a locally unstable manifold.*

In the lectures of A. G. Kušnirenko and A. B. Katok contained in this volume only mappings defined everywhere are considered. However there is no difficulty in extending the definition of a hyperbolic invariant set to the case of local maps, leaving in the definitions only those iterations of the mapping S and its differential which make sense. Only the stable and unstable manifolds demand caution. Only for $n_2 = +\infty$ can one define $W^s(p)$ as the set of points q whose trajectories are asymptotically close to the trajectory of p:

$$W^s(p) = \{q;\, \rho(S^n p, S^n q) \longrightarrow 0 \text{ as } n \longrightarrow +\infty\},$$

and $W^u(p)$ is defined analogously by this device only if $n_1 = -\infty$.

Theorem 2 makes it possible to construct points whose trajectories fall into l_m, l_v^- and l_v^+ in a preassigned manner. Recalling the definition of the mapping S, we obtain a rich stock of solutions of (1). Having been given an infinite sequence $w = [m_n]$ or one of the sequences (23), we can with the aid of the mapping φ find initial data $(v, \tau) = \varphi(w)$ defining a solution $x(t)$ for which the distances between successive zeros are close to $2\pi m_n$, and if the sequence is bounded, then on the corresponding side the solution tends to infinity and the boundary symbol V^- or V^+ gives its limiting velocity. To a periodic sequence there correspond periodic solutions of (1); it is clear that we thus find a countable number of them. The periodic solutions referred to in Theorem 1 correspond to the simplest sequences, where all the m_n are equal.

One can generalize Theorem 2 to the case when the curves Π_0^+ and Π_0^- have several points of transversal intersection. A statement of the corresponding assertions may be found in [59], Part III, and [62]. In the next section we consider a special equation for which Π_0^+ and Π_0^- consist of two points. [62] also gives an axiomatic description of S, to which all the above constructions can be applied.

We pause briefly once more to discuss the case of a well of infinite depth $J = \infty$. For a typical equation one may construct a special neighborhood U, extended along a ray $\tau = \tau_0$, where τ_0 is a simple root of the equation

$$\psi(\tau) = \int_0^\infty xQ'_t(x, \tau)dx = 0.$$

The intersection $U \cap S^{-1}U$ again contains a countable number of "regular" components l_m, but now Sl_m no longer intersects every $l_{m'}$, as was the case if $J < \infty$ (Figure 30), but only those for which the index m' is close to m (Figure 31). Introducing the space of symbols

$$\Omega(d(m)) = \{w = [m_n];\ |m_{n+1} - m_n| < \varphi(m_n)\}, \tag{26}$$

one may prove an analogue of Theorem 2. The condition

$$Q_0(x)\int_0^x Q_0(y)dy \longrightarrow 0,$$

present already in the statement of Theorem 1, here ensures the nontriviality of the result. It means that the distance between fixed points of S lying near the ray $\tau = \tau_0$ diminishes as $v \longrightarrow \infty$. Hence Sl_m intersects a greater and greater number of components $l_{m'}$ if m increases, and $d(m) \uparrow \infty$ in (26).

The differential properties of S, including the property of typicality, are closely connected with the variational equation formed for the equation (1):

$$\delta\ddot{x} = -Q'_x(x_P(t), t)\,\delta x. \tag{27}$$

Let $J < \infty$, and let $x_P(t)$ be a solution generated by the initial data $P = (v, \tau) \in \Pi_0^+ \cap \Pi_0^-$ (see above), parabolic on both sides. From condition C, which must be satisfied by equation (1), it follows that for large $|t|$ the coefficient on the right-hand side of (27) is positive. This makes it possible to prove the existence of two solutions of (27) (we denote them by $\delta_1 x$ and $\delta_2 x$) such that $\delta_1 x(t) \downarrow 0$ as $t \longrightarrow +\infty$ and $\delta_2 x(t) \downarrow 0$ as $t \longrightarrow -\infty$. The equation (1) is typical if $\delta_1 x$ and $\delta_2 x$ are linearly independent (the values of $\delta_1\dot{x}(\tau)$ and $\delta_2\dot{x}(\tau)$ determine the direction of tangents in $\Pi_0^\pm$).

We remark that in the example from the last section the $\delta_i x$ have a power (and not an exponential) order of decay at infinity.

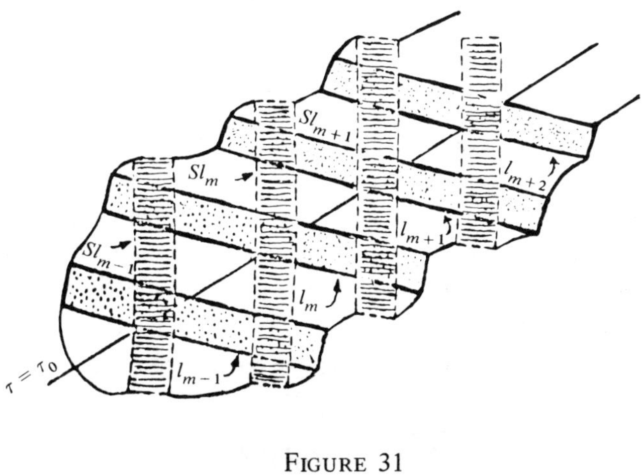

FIGURE 31

§3. On a particular case of the three-body problem

We consider once more, as in Part I, the motion of three bodies (material points) P_i, $i = 1, 2, 3$, acting on one another according to Newton's law; μ_i is the mass of P_i.

We assume that $\mu_1 = \mu_2$ and that at the initial moment the bodies P_1 and P_2 and their velocities are symmetrical with respect to $0Z$, and that the body P_3 lies on this axis and its velocity is directed along it (Figure 32). By considerations of symmetry, a similar distribution of the bodies P_1, P_2 and P_3 will hold for all t, which permits a

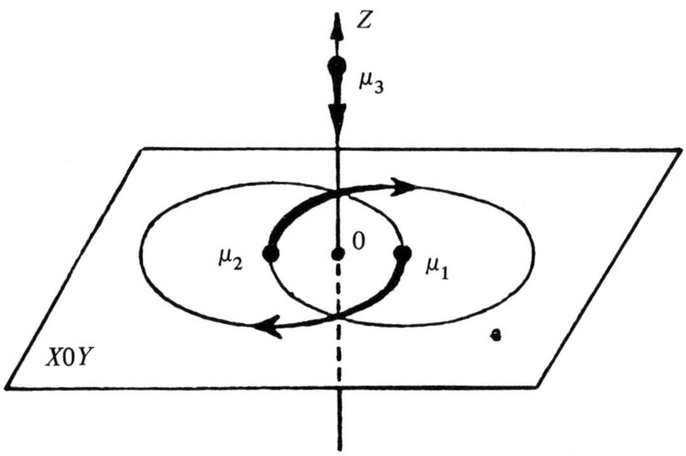

FIGURE 32

reduction of the investigation to a system with two degrees of freedom.

This particular case of the three-body problem was invented long ago (I know a reference to a work of 1895), but the happy idea of investigating the distribution in phase space of trajectories of various final types is due to Kolmogorov. It was precisely for this example that Sitnikov succeeded in proving [29] the existence of oscillating solutions (see Part I), and the author proved the possibility of applying the methods of "symbolic dynamics".

Let $\mu_3 = 0$ (the assumption $0 < \mu_3 \ll 1$ does not introduce any essential novelty). If the usual convention regarding the center of gravity of the system (formula (2) in Part I holds, then P_1 and P_2 describe Keplerian orbits in the plane $X0Y$, symmetric with respect to 0; in the case $h < 0$ and $C \neq 0$ they will be ellipses; we assume this in the sequel. By a suitable choice of units we may take it that the period of rotation of P_1 and P_2 in these ellipses is equal to 2π, and the gravitational constant and the total mass of P_1 and P_2 is equal to unity. We denote by $r(t)$ half of the distance r_{12} (i.e. the distance from each of the bodies P_1, P_2 to the point 0). Then the equation of motion of P_3, whose position is determined solely by the one coordinate z, has the form

$$\ddot{x} = -\frac{z}{(z^2 + r(t)^2)^{3/2}}. \qquad (28)$$

This equation belongs to the class considered in §1; it is easy to verify that its right-hand side satisfies assumptions A–F with $J < \infty$.

The polar coordinates (v, τ), which played an important role in §§1 and 2, are now interpreted as follows. At the instant when P_3 passes through 0 and all three bodies are on one line ("rectilinear configuration"), the state of the system is determined by two quantities: the velocity of P_3 and the phase τ of the elliptic motion of P_1 and P_2 (as τ one may take the mean or true anomaly). By considerations of symmetry, we may confine attention to the modulus of the velocity v of P_3. Then by giving the polar coordinates (v, τ) of a point in the plane Φ, the motion is fully determined to within the direction of the velocity.

If the orbits of P_1 and P_2 are circular, then (in view of the convention regarding the choice of units) $r(t) \equiv ½$ and the equation (28) can be integrated. Initial data (v, τ) belonging to the circle $v = 2$ generate parabolic motions, for $v > 2$ hyperbolic, and for $v < 2$ bounded (the motion of P_3 will then be periodic, just as the motion of the pair P_1, P_2, though the periods of these two motions will be incommensurable on a set of total measure).

In the case when P_1 and P_2 rotate about 0 in ellipses, the curves of parabolic initial data Π_0^+ and Π_0^- do not coincide with one another. If the ellipses differ little from circles, one can prove that the equation is "typical". Using expansions in the eccentricity of the ellipses as a small parameter, one may see that the curves Π_0^+ and Π_0^- are asymptotically close to the circle $v = 2$, and have precisely two transversal points of intersection (as in Figure 3). To one point of intersection, close to $(2, 0)$, there correspond instants of closest approach of the bodies P_1 and P_2, and to the other, close to $(2, \pi)$, instants of greatest separation. Following the arguments of the previous section, one can define neighborhoods of points of intersection that are "good" from the point of view of the possibility of using symbolic dynamics. In phase space, to these neighborhoods there corresponds a certain open subset V of the manifold of rectilinear configurations. It depends on two parameters N and δ; the theorem formulated below is valid if δ is sufficiently small and N sufficiently large.

THEOREM 3. *The set M_V of solutions of the above particular case of the three-body problem, when the instants of the rectilinear configurations of their state belong to V, are in one-to-one correspondence with the set of all symbolic sequences of the form*

$$v^- i_{n_1} \,\vdots\, \cdots m_{n-1} i_{n-1} m_n i_n m_{n+1} i_{n+1} \cdots \,\vdots\, v^+, \tag{29}$$

where m_n, $n_1 < n < n_2$, are natural numbers $\geqslant N$, $i_n = 0$ or 1, $v^\pm \in [0, \delta]$ and $n_1 < 0 \leqslant n_2$; the parts of (29) separated off by dashed lines may be absent ($n_1 = -\infty$ or $n_2 = +\infty$), and then the sequence on the corresponding side is infinite.

The number of rectilinear configurations in the solution is determined by the symbols i_n in (29): if $i_n = 0$, then the $(n-2)$th passage of P_3 through 0 takes place near a moment of closest approach of P_1 and P_2; if $i_n = 1$, then near a moment of greatest separation. The symbol m_n, standing between i_{n-1} and i_n, indicates the number of complete revolutions of P_1 and P_2 around 0 between the $(n-1)$th and nth return of P_3 to 0. The symbols v^- or v^+, if they are present, indicate that P_3 goes to infinity as $t \longrightarrow -\infty$ or $+\infty$ with velocities v^- or v^+, respectively.

The term "solution" in this theorem means a trajectory in the phase space, in which as the origin we take one of the states with rectilinear configuration (it follows from §1 that for every trajectory such states can be found). To this state is assigned the number 0, and to it there corresponds the symbol i_0 in the sequence (29); other states with rectilinear configuration are numbered from zero in both directions. The choice of another rectilinear configuration as the origin leads to a change in the numbering of the symbols in the sequence (29), but the sequence itself remains the same. The final type to which a motion belongs, in the sense of Chazy, is determined by the sequence (29) as in §2, except that the classes $\mathrm{H}^\pm$ and $\mathrm{P}^\pm$ in the sense of §1 are here replaced by $\mathrm{HE}_3^\pm$ and $\mathrm{PE}_3^\pm$.

If the mass μ_3 of the body P_3 is different from zero, the problem becomes more complicated, and results based on a study of (1) cannot be immediately applied to it. Nevertheless, in this case too one succeeds, at least for $\mu_3 \ll \mu_1 = \mu_2$ (see [63]), in constructing the mapping S with the same properties as in §2. This justifies the following corollary,

COROLLARY. *There exists an* $\epsilon > 0$ *such that for* $\epsilon m_1 = \epsilon m_2 > m_3 \geqslant 0$ *and for* $h < 0$ *in the above special case of the three-body problem all* 16 *logically possible combinations of final types of the motion in the sense of Chazy are realized.*

For example, every sequence $v i_0 m_1 i_1 m_2 i_2 m_3 i_3 \ldots$, in which $v > 0$ and $\sup_{n>0} m_n < \infty$, generates a solution for which there occurs complete capture (Part I, §5). By standard devices of symbolic dynamics one can also construct other examples of a motion with preassigned behavior of various forms. Thus, to the sequence

$$\underbrace{v^-0N_1 1N_1 0 \cdots 1N_1}_{2k_1 \text{ times}} \qquad \underbrace{0N_2 0 \cdots N_2}_{2k_2 \text{ times}} 0N_1 1N_1 0N_1 1 v^+$$

there corresponds a motion of a body, which may be interpreted as follows. A space ship, having flown into a double star system from infinity with velocity v^-, left at first on a nearly periodic orbit (of pendulum type) with period $4\pi N_1$. On this orbit it executed k_1 complete oscillations, returning twice during each oscillation to the point of relative equilibrium (the center of mass of the double star), in turn at the moments of closest approach and greatest separation of the components of the double star. The space ship then transferred to an orbit nearly periodic with period $4\pi N_2$, executed k_2 complete oscillations on it, wherein each of its $2k_2$ returns to the point of equilibrium took place near moments of approach of the components of the double star. Finally, returning to the original orbit, the space ship carried out one and a half oscillations about it and departed for infinity in the same direction whence it came. It is remarkable that the whole of this evolution is possible solely on the basis of the force of gravitation, and does not require the expenditure of fuel. True, the precision with which the initial data are given must be extremely high.

Theorem 3 and its corollary made it possible to obtain an answer to a series of theoretical questions of celestial mechanics, among which the question of total capture has probably the greatest interest. None the less, a whole series of problems, as already mentioned in Part I, remain unsolved. For the present example, one has in the first place the question of the measure of the set $OS^{\pm}$ of oscillating solutions. Apart from that, it would be extremely interesting to extend these results to the case of a "full-measure" phase space of the three-body problem, to say nothing of the general problem of n bodies.(6)

BIBLIOGRAPHY

1. George D. Birkhoff, *Dynamical systems,* Amer. Math. Soc., Providence, R. I., 1927.

2. I. M. Gel'fand and S. V. Fomin, *Calculus of variations,* Fizmatgiz, Moscow, 1961; English transl., Prentice-Hall, Englewood Cliffs, N. J., 1963.

3. G. N. Dubošin, *Celestial mechanics: Analytical and qualitative methods,* "Nauka", Moscow, 1964. (Russian)

4. L. D. Landau and E. M. Lifšic, *Mechanics,* Fizmatgiz, Moscow, 1958; English transl., *Course of theoretical physics.* Vol. 1, Pergamon Press, Oxford; Addison-Wesley, Reading, Mass., 1960.

5. Carl Ludwig Siegel, *Vorlesungen über Himmelsmechanik,* Springer-Verlag, Berlin, 1956.

6. Carl Ludwig Siegel and Jürgen K. Moser, *Lectures on celestial mechanics,* Springer-Verlag, Berlin and New York, 1971.

7. J. Milnor, *Morse theory,* Princeton Univ. Press, Princeton, N. J., 1963.

8. Henri Poincaré, *Selected works.* Vols. 1, 2, "Nauka", Moscow, 1971, 1972. (Russian)*

9. ———, *Les méthodes nouvelles de la méchanique céleste.* Tomes I, II, III, Gauthier-Villars, Paris, 1892, 1893, 1899; reprint, Dover, New York, 1957; English transl., Clearinghouse for Federal Sci. Tech. Information, Springfield, Va., 1967.

10. P. K. Raševskiĭ, *Riemannian geometry and tensor analysis,* 3rd ed., "Nauka", Moscow, 1967; German transl. of 1st ed., VEB Deutscher Verlag, Berlin, 1959.

11. Aurel Wintner, *The analytical foundations of celestial mechanics,* Princeton Univ. Press, Princeton, N.J., 1941.

12. E. T. Whittaker, *A treatise on the analytical dynamics of particles and rigid bodies,* 4th ed., Cambridge Univ. Press, New York, 1937; reprint, Dover, New York, 1944.

13. G. F. Hil'mi, *The n-body problem in celestial mechanics and cosmogony,* Izdat. Akad. Nauk SSSR, Moscow, 1951. (Russian)

14. Carl Ludwig Charlier, *Die Mechanik des Himmels,* 2nd rev. ed., vols. 1, 2, de Gruyter, Berlin, 1927.

15. V. M. Alekseev, *Exchange and capture in the three-body problem,* Dokl. Akad. Nauk SSSR **108** (1956), 599–602. (Russian)

16. ———, *Final motion in the three-body problem,* Problems of Motion of Artificial Celestial Bodies (Conf. General and Appl. Questions Theoret. Astronom., Moscow, 1961), M. F. Subbotin et al., editors, Izdat. Akad. Nauk SSSR, Moscow, 1963, pp. 50–64; English transl., No. FTD-MT-64-226, Transl. Div., Wright-Patterson AFB, Dayton, Ohio, 1964.

(6) In Chapters III and VI of Moser's new book [71] there is a complete proof of Theorem 3 which is simpler than that in [59]. One can also find there results supplementing the exposition of Part II.

Editor's note.* The "Celestial Mechanics" section of these volumes consists of Russian translations of [9] and of Poincaré's two papers, *Sur le problème de trois corps et les équations de la dynamique,* Acta Math. **13 (1890), 1–270 (*Oeuvres,* Vol. 7, pp. 262–479), and *Le problème de trois corps,* Rev. Générale Sci. Pures Appl. **2** (1891), 1–5 (*Oeuvres,* Vol. 8, pp. 529–537).

17. D. V. Anosov and Ja. G. Sinaĭ, *Certain smooth ergodic systems,* Uspehi Mat. Nauk **22** (1967), no. 5 (137), 107–172; English transl. in Russian Math. Surveys **22** (1967).

18. V. I. Arnol′d, *Small denominators and problems of stability of motion in classical and celestial mechanics,* Uspehi Mat. Nauk **18** (1963), no. 6 (114), 91–192; English transl. in Russian Math. Surveys **18** (1963).

19. G. A. Krasinskiĭ, *Quasiperiodic solutions of the first sort in the plane n-body problem,* Trudy Inst. Teoret. Astronom. vyp. 13 (1969), 105–168. (Russian)

20. J. E. Littlewood, *A mathematician's miscellany,* Methuen, London, 1953.

21. G. A. Merman, *New criteria of hyperbolic and hyperbolic-elliptic motion in the three-body problem,* Astronom. Ž. **30** (1953), 332–339. (Russian)

22. ———, *On Chazy's investigations in the three-body problem,* Bjull. Inst. Teoret. Astronom. **5** (1951/54), 594–605 (1954). (Russian)

23. ———, *Qualitative investigations of the three-body problem,* Bjull. Inst. Teoret. Astronom. **6** (1955/58), 687–712 (1958). (Russian)

24. G. A. Merman and N. G. Kočina, *Application of a method of estimates to O. Ju. Šmidt's example,* Bjull. Inst. Teoret. Astronom. **6** (1955/58), 85–111 (1955). (Russian)

25. Ju. S. Osipov, *Geometric interpretation of Kepler's problem,* Uspehi Mat. Nauk **27** (1972), no. 2 (164), 161. (Russian)

26. V. F. Proskurin, *On capture in the example of O. Ju. Šmidt,* Bjull. Inst. Teoret. Astronom. **5** (1951/54), 429–434 (1953). (Russian)

27. R. O. Saakjan, *On the probability of capture in the three-body problem,* Izdat. Akad. Akad. Nauk Armjan. SSR, Erevan, 1961. (Russian)

28. K. A. Sitnikov, *On the probability of capture in the three-body problem,* Mat. Sb. **32 (74)** (1953), 693–705. (Russian)

29. ———, *Existence of oscillatory motions in the three-body problem,* Dokl. Akad. Nauk SSSR **133** (1960), 303–306; English transl. in Soviet Phys. Dokl. **5** (1960/61).

30. F. A. Sludskiĭ, *On the many-body problem,* Mat. Sb. **9** (1878/81), 536–545. (Russian)

31. S. Smale, *Topology and mechanics.* I, II, Invent. Math. **10** (1970), 305–331; **11** (1970), 45–64.

32. George Sokoloff [Ju. D. Sokolov], *Conditions d'une collision générale des trois corps, qui s'attirent mutuellement suivant la loi de Newton,* Ukraïn. Akad. Nauk Trudi Fiz.-Mat. Viddilu = Acad. Sci. Ukraïne Mém. Cl. Sci. Phys. Math. **9** (1928/29), no. 1, 1–62. (Ukrainian; French summary)

33. G. F. Hil′mi, *On the possibility of capture in the three-body problem,* Dokl. Akad. Nauk SSSR **62** (1948), 39–42. (Russian)

34. O. Ju. Šmidt, *On the possibility of capture in the three-body problem,* Dokl. Akad. Nauk SSSR **58** (1947), 213–216.

35. V. M. Alexeyev [Alekseev], *Sur l'allure finale du mouvement dans le problème des trois corps,* Proc. Internat. Congr. Math. (Nice, 1970), vol. 2, Gauthier-Villars, Paris, 1971, pp. 893–907.

36. L. Becker, *On capture orbits,* Monthly Notices Roy. Astronom. Soc. **80** (1919/20), 590–597.

37. Jean Chazy, *Sur certaine trajectoires du problème des n corps,* Bull. Astronom. **35** (1918), 321–389.

38. ———, *Sur l'allure du mouvement dans le problème des trois corps quand le temps croit indéfiniment,* Ann. Sci École Norm. Sup. (3) **39** (1922), 29–130.

39. ———, *Sur l'allure finale du mouvement dans le problème des trois corps,* J. Math. Pures Appl. (9) **8** (1929), 353–380.

40. ———, *Sur l'allure finale du mouvement dans le problème des trois corps,* Bull. Astronom. (2) **8** (1932), 403–436.

41. V. Fock [Fok], *Zur Theorie des Wasserstoffatoms,* Z. Phys. **98** (1935/36), 145–154.

42. J. E. Littlewood, *On the problem of n bodies,* Medd. Lunds Univ. Mat. Sem. Supplementband [Marcel Riesz 65th Birthday Volume] (1952), 143–151.

43. Jürgen K. Moser, *Lectures on Hamiltonian systems,* Mem. Amer. Math. Soc. No. 81 (1968), 1–60.

44. ———, *Regularization of Kepler's problem and the averaging method on a manifold,* Comm. Pure Appl. Math. **23** (1970), 609–636.

45. P. Painlevé, *Leçons sur la théorie analytique des équations différentielles, professées à Stockholm,* 1895, Hermann, Paris, 1897.

46. Harry Pollard and Donald G. Saari, *Singularities of the n-body problem.* I, Arch. Rational Mech. Anal. **30** (1968), 263–269.

47. Donald G. Saari, *On oscillatory motions in the problem of three bodies,* Celestial Mech. **1** (1969/70), 343–346.

48. ———, *Expanding gravitational systems,* Trans. Amer. Math. Soc. **156** (1971), 219–240.

49. Hans J. Sperling, *On the real singularities of the N-body problem,* J. Reine Angew. Math. **245** (1970), 15–40.

50. Karl F. Sundman, *Recherches sur le problème des trois corps,* Acta Soc. Sci. Fenn. **34** (1906/07), no. 6.

51. ———, *Mémoire sur le problème des trois corps,* Acta Math. **36** (1912/13), 105–179.

52. Marston Morse and Gustav A. Hedlund, *Symbolic dynamics.* I, II, Amer. J. Math. **60** (1938), 815–866; **62** (1940), 1–40.

53. Norman Levinson, *A second order differential equation with singular solutions,* Ann. of Math. (2) **50** (1949), 127–153.

54. J. E. Littlewood, *On non-linear differential equations of the second order.* III, IV, Acta Math. **97** (1957), 267–308; **98** (1957), 1–110.

55. V. A. Pliss, *Nonlocal problems in the theory of oscillations,* "Nauka", Moscow, 1964; English transl., Academic Press, New York, 1966.

56. S. Smale, *A structurally stable differentiable homeomorphism with an infinite number of periodic points,* Qualitative Methods in the Theory of Nonlinear Vibrations (Proc. Internat. Sympos. Nonlinear Vibrations (Kiev, 1961), Vol. II), Izdat. Akad. Nauk Ukrain. SSR, Kiev, 1963, pp. 365–366.

57. R. L. Adler and B. Weiss, *Entropy, a complete metric invariant for automorphisms of the torus,* Proc. Nat. Acad. Sci. USA **57** (1967), 1573–1576.

58. V. M. Alekseev, *Perron sets and topological Markov chains,* Uspehi Mat. Nauk **24** (1969), no. 5 (149), 227–228. (Russian)

59. ———, *Quasirandom dynamical systems.* I, II, III, Mat. Sb. **76 (118)** (1968), 72–134; **77 (119)** (1968), 541–601; **78 (120)** (1969), 3–50; English transl. in Math. USSR Sb. **5** (1968); **6** (1968); **7** (1969).

60. Stephen Smale, *Diffeomorphisms with many periodic points,* Differential and Combinatorial Topology (A Symposium in Honor of Marston Morse), Princeton Univ. Press, Princeton, N. J., 1965, pp. 63–80.

61. L. P. Šil'nikov, *On a Poincaré-Birkhoff problem,* Mat. Sb. **74 (116)** (1967), 378–397; English transl. in Math. USSR Sb. **3** (1967).

62. V. M. Alekseev, *Quasirandom oscillations,* Proc. Fifth Internat. Conf. Nonlinear Oscillations (Kiev, 1969), Vol. 2, Izdanie Inst. Mat. Akad. Nauk Ukrain. SSR, Kiev, 1970, pp. 33–38. (Russian)

63. ———, *Quasirandom dynamical systems,* Doctoral Dissertation, Moscow State Univ., Moscow, 1969; abstract published in Mat. Zametki **6** (1969), 489–498; English transl. in Math. Notes **6** (1969).

64. V. I. Arnol'd, *Lectures on classical mechanics,* Izdat. Moskov. Gos. Univ., Moscow, 1968. (Russian)

65. Eberhard Hopf, *Ergodentheorie,* Springer-Verlag, Berlin, 1937; reprint, Chelsea, New York, 1948.

66. V. V. Nemyckiĭ and V. V. Stepanov, *Qualitative theory of differential equaitons,* 2nd ed., GITTL, Moscow, 1949; English transl., Princeton Univ. Press, Princeton, N. J., 1960.

67. Joseph Louis Lagrange, *Essai sur le problème des trois corps,* Prix Acad. Roy. Sci. Paris **9** (1772); reprinted in his *Oeuvres,* Vol. 6, Gauthier-Villars, Paris, 1873.

68. Leonhard Euler, *Theoria motuum lunae nova methodo pertractata,* Typis Acad. Imp. Sci., St. Petersburg, 1772; reprinted in his *Opera omnia.* Ser. II, Vol. 22, Soc. Sci. Nat. Helv., Lausanne, and Orell Füssli, Zurich, 1958.

69. Ju. I. Neĭmark, *Motions close to a doubly asymptotic motion,* Dokl. Akad. Nauk SSSR **172** (1967), 1021–1024; English transl. in Soviet Math. Dokl. **8** (1967).

70. ———, *Structure of motions of a dynamical system in the neighborhood of a homoclinic curve,* Fifth Math. Summer School (Uzhgorod, 1967), Akad. Nauk Ukrain. SSR Inst. Mat., Kiev, 1968, pp. 400–435. (Russian)

71. Jürgen Moser, *Stable and random motions in dynamical systems,* Ann. of Math. Studies, No. 77, Princeton Univ. Press, Princeton, N. J., and Univ. of Tokyo Press, Tokyo, 1973.

Translated by F. V. ATKINSON

ABCDEFGHIJ–CM–8987654321